The Conceptual Foundations of the Statistical Approach in Mechanics

By Paul and Tatiana Ehrenfest

TRANSLATED BY

Michael J. Moravcsik

Dover Publications, Inc.
New York

Published in Canada by General Publishing Company, Ltd., 30 Lesmill Road, Don Mills, Toronto, Ontario.

Published in the United Kingdom by Constable and Company, Ltd., 10 Orange Street, London WC2H 7EG.

This Dover edition, first published in 1990, is an unabridged republication of the English translation first published by The Cornell University Press, Ithaca, NY, in 1959. The pagination of the frontmatter has been reorganized for reasons of space, but nothing has been omitted. The work was first published in German by B. G. Teubner (Leipzig, 1912) as No. 6 of Volume IV:2:II of The *Encyklopädie Der Mathematischen Wissenschaften.*

Manufactured in the United States of America
Dover Publications, Inc., 31 East 2nd Street, Mineola, N.Y. 11501

Library of Congress Cataloging-in-Publication Data

Ehrenfest, Paul, 1880–1933.
[Begriffliche Grundlagen der statistischen Auffassung in der Mechanik. English]
The conceptual foundations of the statistical approach in mechanics / by Paul and Tatiana Ehrenfest ; translated by Michael J. Moravcsik.
p. cm.
Translation of: Begriffliche Grundlagen der statistischen Auffassung in der Mechanik.
"Unabridged republication of the English translation first published by The Cornell University Press, Ithaca, NY, in 1959"—T.p. verso.
Includes bibliographical references.
ISBN 0-486-66250-0
1. Gases, Kinetic theory of. 2. Statistical mechanics.
I. Ehrenfest, Tatiana, 1876- . II. Title.
QC175.E353 1990
530.1′3—dc20 89-77772
CIP

Foreword

IN recent years there has been a strong revival of interest in the foundations of statistical mechanics, and a great deal of important work has been done both in this country and abroad.

"The Conceptual Foundations of Statistical Mechanics" is the title of a celebrated article which the late Paul Ehrenfest prepared in collaboration with his wife Tatiana Ehrenfest for the German *Encyclopedia of Mathematical Sciences.* The article appeared in 1912 and has since become a classic. In spite of the passage of time it has lost little of its scientific and didactic value, and no serious student of statistical mechanics can afford to remain ignorant of this great work.

Unfortunately it is not readily accessible and it requires a knowledge of German far beyond that of an average reader. It seemed therefore appropriate to make the article available in English, and the present translation, by Dr. Michael J. Moravcsik, is being offered to the public.

The translation adheres as rigidly as possible to the original. Because of this, the reader may be somewhat mystified by numbers like V 8 or VI(2) 21 which occasionally appear before references. These refer to other volumes of the *Encyclopedia* in which related articles have appeared. The bibliography at the end of the book is taken verbatim from the original, and no attempt has been made to bring it up to date. In this edition, the original notes are all collected at the end of the text. They are extremely important and contain a great deal of wisdom and wit, though the latter could not always be reflected in translation. A number of German words proved to be difficult to translate without sacrifice of nuances of meaning. They were left intact but italicized in the text and explained in footnotes.

It is hoped that the translation will be of help to student and teacher alike and that it will stimulate further work in this important and fascinating field.

M. Kac
G. E. Uhlenbeck

June 1959

Authors' Preface

THIS article is closely related to V 8 by L. Boltzmann and J. Nabl ("Kinetische Theorie der Materie"). Both articles deal with the application of methods of probability theory in investigations of the motion of systems of molecules. However, whereas V 8 is concerned mainly with the physical results, the present work will take up the conceptual foundations of the procedure.

Since 1876 numerous papers have called attention to these foundations. In these papers the Boltzmann H-theorem, a central theorem of the kinetic theory of gases, was attacked. Without exception all studies so far published dealing with the connection of mechanics with probability theory grew out of the synthesis of these polemics and of Boltzmann's replies. These discussions will therefore be referred to frequently in our report.

For the connections with V 3 by G. W. Bryan ("Allgemeine Grundlagen der Thermodynamik"), with V 23 by W. Wien ("Theorie der Strahlung"), and with VI(2) 21 by S. Oppenheim ("Figur des Saturnringes"), see the last paragraphs of this article.

Preface to the Translation

THE great task of collecting the literature and of organizing the *Encyklopädie* article was done by Paul Ehrenfest. My contribution consisted only in discussing with him all the problems involved, and I feel that I succeeded in clarifying some concepts that were often incorrectly used. To this English translation of the *Encyklopädie* article, I would like to add the following remarks, which, in my opinion, should have been included in the original version of the article and for the omission of which I feel personally responsible.

1. At the time the article was written, most physicists were still under the spell of the derivation by Clausius of the second law of thermodynamics in the form of the existence of an integrating factor for the well-known expression for the quantity of heat ΔQ put into the system. In this derivation the irreversibility in time of all processes occurring in nature played an important role. Hence it seemed that the possibility of a reversal of the natural development (which according to the *Wiederkehreinwand* of Zermelo should occur after a sufficiently long time) threatened the validity of some of the most important results of thermodynamics. However, it became clear to me afterwards, that the existence of an integrating factor has to do only with the mathematical expression of $\Delta Q = dU + dA$ in terms of the differentials dx_1, dx_2, $\cdots$, dx_n of the *equilibrium* parameters x_1,

$x_2, \cdots, x_n$ and is completely independent of the direction in time of the development of the natural processes.[1] As a result, the fact of the reversibility of the mechanical motion, which is inescapable in the kinetic interpretation of the laws of thermodynamics, lost some of its importance. Nevertheless, even today many physicists are still following Clausius, and for them the second law of thermodynamics is still identical with the statement that the entropy can only increase.[2]

2. As a result of the above-mentioned point of view, the attention of most physicists was concentrated on the problem of the *change* of the H-function in time. Although Boltzmann had to admit that his *Stosszahlansatz*, on which the calculation of the change of H was founded, could not always be valid, he insisted that the probability of deviations was very small. This led to many efforts on the part of physicists to prove that one would always be more likely to find a system in the state of decreasing H than in the state of increasing H. The article of Paul and Tatiana Ehrenfest, "Über zwei bekannte Einwande gegen das Boltzmannsche H-Theorem" (*Phys. Zeits.*, 1907), is in a way one of these attempts; it tries to show how one could reconcile the high probability of the validity of the *Stosszahlansatz* with the inescapable quasi-periodic recurrence of the same value of H.

3. It turned out that many physicists often confused two essentially different problems: (a) the relative probabilities of a decrease and of an increase of the H-function starting from a *given* value of H (which is different from the minimum value) and (b) the relative probabilities of a transition of the H-function from a higher to a lower value and of a transition in the opposite direction. The difference between these two problems I discussed in a paper, "On a Misconception in the Probability Theory of Irreversible Processes" (*Proc. Amst. Acad.*, vol. 38, 1925).

4. Since for the periods of increase the *Stosszahlansatz*

cannot be valid, the question arises in which sense one should take the assertion of Boltzmann that deviations from the *Stosszahlansatz* are very improbable. As I see it now, the answer must be that for overwhelmingly long times when the H-function *stays* at its minimum value, neither of the objections against the H-theorems apply, and hence the *Stosszahlansatz* should be applicable, which explains the high probability of its occurrence. In this way the ingenious derivation of the formula for the change of H, that is, the H-theorem, can be rehabilitated, although by emphasizing another aspect than Boltzmann intended.

5. Although Boltzmann did not fully succeed in proving the tendency of the world to go to a final equilibrium state, there remain after all criticisms the following valuable results: first, the derivation of the Maxwell-Boltzmann distribution for equilibrium states, then the kinetic interpretation of the entropy by the H-function, and finally the *explanation* of the existence of an integrating factor for $dU+dA$. In thermodynamics the existence of such a factor is always based on an unexplained hypothesis.

The very important irreversibility of all observable processes can be fitted into the picture in the following way. The period of time in which we live happens to be a period in which the H-function of the part of the world accessible to observation decreases. This coincidence is really not an accident, since the existence and the functioning of our organisms, as they are now, would not be possible in any other period. To try to explain this coincidence by any kind of probability considerations will, in my opinion, necessarily fail. The expectation that the irreversible behavior will not stop suddenly is in harmony with the mechanical foundations of the kinetic theory.

T. EHRENFEST-AFANASSJEWA

Leiden, The Netherlands
February 1959

Contents

Introduction

1. Background

Older works on the kinetic theory of gases quite uniformly show the following attitude toward the application of probability theory: The goal is the "explanation" of the observable aerodynamic processes on the basis of two groups of "assumptions." These are:

1. Assumptions about the mechanical structure. Each gas quantum is a mechanical system, consisting of an enormous number[1] of identical molecules of strictly specified structure.[2]

2. The so-called "probability assumptions." In the motion of molecules, which is too complicated to be observed, certain regularities are described in terms of statements about the relative frequency of various configurations and motions of the molecules.[3] (Cf. Sections 3–5.)

Of course, the more conceptual analysis and criticism of these foundations attracted no attention at first because of the rich yield of results that could be compared with experiment. These results were obtained in fast succession by A. Krönig (1856), R. Clausius (since 1857), and J. C. Maxwell (1859) through the kinetic interpretation of the equation of state, diffusion, heat conduction, and viscosity.[4]

The incentive for a thorough discussion of the foundations was first given by the Boltzmann H-theorem (1872).[5]

L. Boltzmann came to the following conclusion: The assumptions used by Clausius and Maxwell are sufficient to give a unified interpretation of irreversible phenomena. In particular, they give a kinetic meaning to the monotonic increase of entropy with time.[6]

On the other hand, J. Loschmidt (1876)[7] and later other authors, among them especially E. Zermelo (1876)[8] favored arguments which can be crystallized in the statement: It follows from the basic postulates of kinetic theory that equally large increases and decreases in the entropy are completely *gleichberechtigt*.[*,9]

Thus in the study of the most general problems in kinetic theory one was faced with two results that seemed to be completely irreconcilable with each other. It was plausible to put the blame for this on some internal contradiction in the foundations of the mechanico-probabilistic theory. (Cf. Sections 7 and 16.)

Boltzmann in his replies developed a somewhat modified and more precise formulation of the H-theorem and of his own ideas on the foundations as well as of the arguments of his opponents. Nowadays this formulation is referred to as "statistico-mechanical."[10] Of Boltzmann's conclusions it may be said:

1. In this more rigorous, statistical formulation the apparent contradiction between his results and those of Loschmidt and Zermelo disappears.

2. This modified formulation of the H-theorem also agrees over the entire field of possible observations with the requirements of the monotonic increase of entropy.

However, several workers[11] of good reputation still hold that the objections of Loschmidt and Zermelo reveal inner contradictions in the foundations even in the

* "Being on an equal footing, having equal rights." A few German terms that cannot be translated into English except by cumbersome phrases have been left in the German form.

modified, statistical formulation of the H-theorem.

Our discussion will be guided by the conviction that such inconsistencies do not exist and that wherever they seem to arise they are the result of ambiguities which may be due to certain terms used by Boltzmann (cf., in particular, the apparently self-contradictory geometrical properties of the Boltzmann "H-curves"—Section 14b). Consequently we will also take a stand against the viewpoint that the Clausius-Maxwell interpretation of diffusion, heat conduction, and viscosity must be discarded and that "its apparent success must be attributed to faulty reasoning."[12]

Since the "statistico-mechanical" studies arose from this fight about the H-theorem, they are very far from presenting a systematic treatment. Instead, they should be regarded as a collection of illuminating comments on the older probability assumptions of the theory of gases. They reformulate the older postulates in a more precise form by transforming the terminology of probability theory step by step into hypothetical statements about the relative frequencies in clearly defined statistical ensembles. Certain obscurities, inherent in the probabilistic terminology, were in this way eliminated.[13] The numerous gaps which thus became obvious (cf. "statements" I–X in Sections 15–16) induced further investigations, at least one of which was carried through by Boltzmann with great success (cf. Section 13).

A close look at this process of development shows that a systematic treatment, which W. Gibbs attempts to give in his *Elementary Principles in Statistical Mechanics*, covers only a small fraction of the ideas (and not even the most important ones) which have come into being through this process (cf. Sections 19–25).

For this reason it seemed mandatory to choose an essentially historic approach and to start our discussion with the older formulation of the foundations of the theory.[14]

I

The Older Formulation of Statistico-mechanical Investigations (Kineto-statistics of the Molecule)

2. The first provisional probability postulates[15]

The first quantitative attempts at a kinetic interpretation were concerned with the equation $pv=RT$, i.e., with the calculation of the pressure which the molecules of a gas at rest exert on the wall of the container through their heat motion.[16] These attempts at computing the pressure replace the unknown and unexplorably complicated motion of the molecules with a model of the motion, which is chosen on the basis of convenience in calculation. In this model all molecules move with the same speed, one-third of them moving parallel to each one of the three perpendicular edges of the cubical container. Krönig (1856),[17] for one, justified this method in these words: "The path of each molecule must be so irregular that it will defy all calculation. However, according to the laws of probability theory, one can assume a completely regular motion in place of this completely irregular one."[18]

3. The equal frequency of apparently gleichberechtigt *occurrences*

3a. The assumptions of Clausius. In the works of R. Clausius (since 1857)[19] we can find a spirit substantially more critical both in methods and in form of expression. His assumptions are limited, almost without exception, to statements about the equal frequency of those motions and configurations of molecules which he thinks are plausibly enough *gleichberechtigt.*

Let us take, for instance, a gas at rest, in thermal equilibrium, in the absence of external forces. In this case Clausius, partly tacitly, partly explicitly, bases his calculations on the following statements:

1. The molecules are distributed with equal density throughout the container.
2. The various values of the speed of molecules occur with the same frequencies in the different parts of the container.
3. If we take a volume element which contains a sufficiently large number of molecules, all directions of velocity are represented with equal frequency.[20]

3b. The Stosszahlansatz.*[,21] We must also mention here the extremely important assumption, in essence already given by Clausius, about the number of collisions which take place in a time element Δt between two groups of molecules moving toward each other. One computes the sum of the volumes of the cylinders which the "action spheres"[22] of the molecules from the first group sweep during Δt in their relative motion with respect to the molecules of the second group.[23] The number of collisions is then equal to the number of those molecules from the second group which at time t lie inside these cylinders. For this number Clausius gives the following expression:[24]

[Volume swept through] × [Number of molecules of the second group per unit volume].

This *Stosszahlansatz* is clearly based on the following

* "Assumption about the number of collisions."

assumption about equal frequencies ("equal probabilities"): The number of molecules per unit volume is the same in the space to be traversed as in any other part of the space.[25]

*4. The relative frequency of non-*gleichberechtigt *occurrences*

4a. The qualitative assumptions and the first estimates of Clausius. On the basis of the assumptions about frequencies discussed above, Clausius gave a new derivation of the equation $pv=RT$ and developed the first quantitative estimate for the diffusion velocity.[26]

At that time, however, Clausius had already expounded at least qualitatively the kinetic meaning of those phenomena whose quantitative treatment calls for much deeper assumptions about frequencies. Thus he showed, for instance,[27] that the equilibrium between a liquid and its saturated vapor at different temperatures depends on the fraction of the molecules in the liquid whose speed exceeds, at the temperature in question, the critical speed necessary to escape from the liquid. Therefore, in order to discuss the equilibrium in a quantitative manner, i.e., as a function of temperature, we need an assumption about the relative frequency of the various molecular speeds.

Clausius, however, makes no attempt to give a quantitative estimate for the relative frequencies of obviously non-*gleichberechtigt* occurrences. As a consequence, in some cases he gives up the quantitative approach altogether.[28] In others he is satisfied with estimates: he replaces the unknown relative frequency by an assumption which is intentionally schematic and which is chosen for convenience in calculation. He emphasizes, however, that he is presenting only an estimate. In his treatment of the velocity of diffusion, for instance, he carries through his calculations as if the molecules move in all directions, but all with the same speed.[29] Even this estimate, however, is based on a somewhat deeper assumption, without which we could not expect these estimates to give even an approximate result. This assumption is that, in a gas at rest at a given temperature, the molecules have a well-defined (even if unknown) distribution of velocity—and, in fact, such a distribution that the speeds show a relatively small dispersion about the most frequent value.[30]

4b. Maxwell's derivation of a law for the distribution of velocities. In order to elaborate further on the conclusions reached by Clausius, it became necessary to convert the qualitative statement about the smallness of the dispersion of the velocity distribution into some specific quantitative assumptions, which could then be used in calculations. This is where J. C. Maxwell entered the scene (1859).[31] Considering the case of a monatomic gas at rest,[32] in thermal equilibrium and in the absence of external forces, he postulated the following law for the distribution of velocities (the "Maxwell distribution law"):

$$f(u, v, w)\Delta u\Delta v\Delta w = Ae^{-B(u^2+v^2+w^2)}\Delta u\Delta v\Delta w. \tag{1}$$

Here $f(u, v, w)\Delta u\Delta v\Delta w$ is the number of molecules, the three velocity components of which are between the limits

(2a) u and $u + \Delta u$

(2b) v and $v + \Delta v$

(2c) w and $w + \Delta w$ respectively.

A and B are two constants which can be determined from the total number, the total mass, and the total kinetic energy of the molecules.[33]

We may assume that Maxwell was guided by the idea

of the Gaussian distribution of error when he stated this distribution law.[34] See Sections 15–21 of V 8 for the central position immediately assumed by the Maxwell hypothesis in the calculation of coefficients of diffusion, heat conduction, and friction.

4c. Boltzmann's generalization of the Maxwell distribution law. Boltzmann (1868)[35] introduced into the theory of gases a significant generalization of Maxwell's distribution law. First of all, he took into account an external field of force acting on the molecules of the gas (e.g., gravity). In this case not only the different speeds (i.e., different values of the kinetic energy) fail to be *gleichberechtigt*, but the different positions of a molecule in the container (i.e., the different values of the external potential energy) fail likewise.[36] Secondly, Boltzmann took into account the fact that each molecule consists of several atoms that are connected with each other by attractive forces.[37] In this case even the various configurations of the atoms in the molecule (i.e., the different values of the internal potential energy) must be considered non-*gleichberechtigt.*

For this very general case Boltzmann's generalization of Maxwell's law states: If $\Delta\tau$ denotes a very small range of variation in the state of a molecule—so characterized that the coordinates and velocities of all atoms are enclosed by suitable limits,[38] similar to limits (2a), (2b), (2c)—then for the case of thermal equilibrium

$$f \cdot \Delta\tau = \alpha e^{-\beta\epsilon} \cdot \Delta\tau \tag{3}$$

gives the number of those molecules whose states lie in the range of variation $\Delta\tau$. Here ϵ denotes the total energy the molecule has in this state (kinetic energy + external potential energy + internal potential energy),[39] and α and β are two constants which are to be determined just as in the case of Maxwell's law. Thus for a spherical molecule of mass m in the presence of a gravitational field, we have

$$\Delta\tau = \Delta x \Delta y \Delta z \Delta u \Delta v \Delta w \tag{3'}$$

$$f = \alpha e^{-\beta((m/2)c^2 + mgz)} \tag{3''}$$

where

$$c^2 = u^2 + v^2 + w^2. \tag{3'''}$$

A comparison with Eq. (1) shows that, as far as the velocity distribution is concerned, Eq. (3″) agrees with the Maxwell assumption. Hence we call Eq. (3) the Maxwell-Boltzmann distribution law.[40]

5. Attempts to derive frequency postulates of the second kind from those of the first kind

The frequency postulates of Maxwell and Boltzmann just presented obviously need a justification even more than do the postulates about equal frequencies that we discussed under Section 4a. In fact, Maxwell and Boltzmann published their laws of distribution not as a hypothesis[41] but as final results of systematic derivations.

Maxwell outlined this derivation in his first paper on the theory of gases;[42] later, however, he discarded it as unsatisfactory. (For the presentation and criticism of this derivation, see V 8, No. 7.)

The subsequent attempts of Maxwell and Boltzmann to obtain a derivation reached their first conclusion in the H-theorem. Before reaching this they had gone through the following stages of development:

Maxwell (1866)[43]—The Maxwell distribution for a monatomic gas, in the absence of external forces, satisfies the requirements of thermal equilibrium,[44] i.e., the molecular collisions do not affect the invariance of the velocity distribution.[45]

Boltzmann (1868–1871)[46]—In polyatomic gases and in the presence of external force acting on the molecules the Maxwell-Boltzmann distribution is still stationary with respect to molecular collisions.

Boltzmann (1872, as one of the corollaries of the H-theorem)[47]—The Maxwell-Boltzmann distribution is the only distribution which can maintain its invariance,[48] and any other distribution under the influence of collisions finally goes over into the Maxwellian one.

We want to emphasize the following features with regard to the foundations of these investigations:

1. The calculation uses, at least partially, the mechanical properties of the gas model. In particular it uses the laws for the collisions of two molecules in order to calculate, when the type of collision is specified, the domains of state ($\Delta\tau$) into which two molecules are thrown from given domains of state.

2. The calculation employs some of the postulates about equal frequencies which are discussed in Section 3. In particular, when the number of collisions of various kinds in time element Δt is sought, it uses an assumption which is essentially identical with the *Stosszahlansatz* discussed in Section 3.

As to the frequency postulates, which lie at the basis of the kinetic theory of gases, the above investigations give the following results: With a partial application of the mechanical properties of the gas model, they prove statements about the relative frequencies of non-*gleichberechtigt* occurrences (the Maxwell-Boltzmann distribution law). In doing so they take for granted in the calculations certain assumptions about equal frequencies (especially the *Stosszahlansatz*).

Thus the *Stosszahlansatz* assumes a central position. Criticism of this postulate and revision of its corollaries will be treated later.[49]

Appendix to Section 5

Because of certain later discussions it seems advisable to explain on a much-simplified model what the position of the *Stosszahlansatz* is in the Maxwell-Boltzmann investigations.

Let us consider a large but finite number (N) of material points moving in the infinite plane. We will call these points the "P-molecules." We will assume that they can completely penetrate each other. They move in the absence of forces, except for elastic collisions that they undergo with the "Q-molecules." The "Q-molecules" are defined as squares with sides of length a; there are an infinite number of them, distributed irregularly over the infinite plane, and they are fixed. Every portion of the plane contains about the same number of them (i.e., the distribution is uniform over the plane), and the average distance A of the neighboring squares is large compared to a. The diagonal of each Q-molecule is exactly parallel to the x and y axes respectively.

We assume that at time t_0 all P-molecules have the same speed c and that they can move in the following four directions:

$$(1) \rightarrow \qquad (2) \uparrow \qquad (3) \leftarrow \qquad (4) \downarrow .$$

Because the Q-molecules are fixed and because their diagonals are oriented exactly, this assumption will hold true at any time. On the other hand, however, the numbers

$$f_1, f_2, f_3, f_4, \tag{4}$$

which are the numbers of molecules moving in the four directions at any given time, will be changed by the collisions of the P-molecules with the Q-molecules. In other words, the "velocity distribution" will change.

The distribution analogous in this example to the Maxwell distribution is

$$\overset{0}{f_1} = \overset{0}{f_2} = \overset{0}{f_3} = \overset{0}{f_4} = \frac{N}{4} \cdot \tag{5}$$

Therefore in our case we have to show that a gradual

equalization of the four f_i's takes place under the influence of the collisions and that the distribution (Eq. 5) maintains itself once it has been achieved.

Let us denote by $N_{12}\Delta t$ the number of P-molecules whose motion is changed by the collision from direction (1) to direction (2). These are all those and only those molecules which at the beginning of time interval Δt satisfy simultaneously the following two requirements:

A. They move in direction (1).

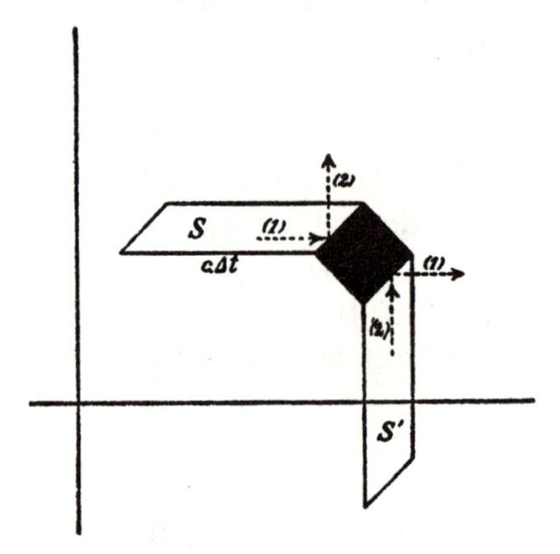

Figure 1

B. They lie in one of the strips S. (See Figure 1; one should imagine that each of the infinitely many Q-molecules has such a strip attached to it.)

It is clear that knowledge of the numbers f_1, f_2, f_3, f_4 is not enough to determine how many P-molecules satisfying condition A will also satisfy condition B.

The analogy to the *Stosszahlansatz* can now be expressed by the following statement:

The fraction of P-molecules of each single direction of motion which lie in the strips S is the same as the ratio of the total area of the strips to the total free area in the plane. Let us denote this ratio by

$$k \cdot \Delta t. \tag{6}$$

Then in time interval Δt

$$N_{12}\Delta t = f_1 \cdot k\Delta t \tag{7}$$

molecules are thrown from (1) to (2); similarly

$$N_{21}\Delta t = f_2 \cdot k\Delta t \tag{8}$$

molecules are thrown in the same time interval from (2) to (1). (Here the strips S are to be replaced by the strips S' which have the same area; see Figure 1.)

A comparison of Eqs. (7) and (8) shows immediately that, because of collisions of the type discussed above, in time interval Δt the larger f loses

$$|f_1 - f_2| \cdot k\Delta t \tag{9}$$

molecules to the smaller f. An analogous statement can be made about every other pair of collision types.

If we use the *Stosszahlansatz* given in Eq. (7) for the calculation of the numbers N_{12}, N_{21}, N_{23}, N_{32}, etc., in each time interval Δt, we get a monotonic decrease in the differences between the numbers f_1, f_2, f_3, f_4. Distribution (Eq. 5) is therefore reached monotonically in time.

6. *The Boltzmann* H*-theorem: The kinetic interpretation of irreversible processes*[50]

Criticism of the *Stosszahlansatz* and its corollaries arose as soon as it was recognized as paradoxical that the completely reversible gas model of the kinetic theory was apparently able to explain irreversible processes, i.e., phenomena whose development shows a definite direction in time. These nonstationary,[51] irreversible processes were brought into the center of interest by the H-theorem of Boltzmann. In order to show that every non-Maxwellian distribution always approaches the Maxwell distribution in time, this theorem synthesizes all the special irreversible processes (like heat conduction and internal friction) which occur at the same time into one total and typical irreversible process. The theorem culminates in the kinetic interpretation of the postulate of thermodynamics that during such irreversible processes the entropy always increases.[52]

In contradistinction to entropy Boltzmann defines a certain one-valued function of the instantaneous state distribution of the molecules, which he calls the H-function.[53] Consider a distribution, which may be arbitrarily different from the Maxwell-Boltzmann distribution, and let us denote by $f \cdot \Delta\tau$ the number of those molecules whose state lies in the small range $\Delta\tau$ of the state variables.[54] Then we define the H-function as

$$H = \sum f \log f \cdot \Delta\tau, \tag{10}$$

where the sum[55] is to be taken over all the possible domains $\Delta\tau$.

The H-theorem shows that because of the collisions the quantity H decreases monotonically with increasing time. If we consider a gas model left to itself, which at times $\cdots t_1, t_2, t_3, \cdots, t_n \cdots$ has reached the phases

$$\cdots \Gamma_1, \Gamma_2, \cdots, \Gamma_{n-1}, \Gamma_n, \cdots \tag{11}$$

of its development, then we have the following inequalities for the corresponding H-values

$$\cdots H_1 \geq H_2 \cdots \geq H_{n-1} \geq H_n \cdots \tag{12}$$

and the equality signs hold when and only when the system has reached the Maxwell-Boltzmann distribution.[56] We see that the quantity H behaves in its time development just as the entropy with a minus sign.[57]

7. *Objections to the result concerning irreversibility*

7a. Loschmidt's Umkehreinwand* (1876).[58] Denote by Γ_s and Γ_s' two phases of the molecular motion for which

* "Objection or counter argument based on reasoning involving reversal in time."

(1) all molecules have the same positions, (2) all molecules in phase Γ_s have velocities equal in magnitude but opposite in direction to the velocities in phase Γ_s'. Then denoting by H_s and H_s' the corresponding values of the H-function, we get from definition (Eq. 10)[59]

$$H_s' = H_s. \tag{13}$$

The gas model is a conservative mechanical system. Therefore if it can perform the motion which has gone through the phases in the order of Eq. (11), it can also perform the motion in the order

$$\cdots \Gamma_n', \Gamma_{n-1}', \cdots, \Gamma_2', \Gamma_1' \cdots. \tag{14}$$

For this motion, using (13) and (12):

$$\cdots H_n' \leq H_{n-1}' \cdots \leq H_2' \leq H_1'. \tag{15}$$

Hence for every motion of the model in which H decreases from H_1 to H_n, there exists a motion in which, in a precisely opposite way, H increases from H_n to H_1.[60]

7b. Zermelo's Wiederkehreinwand.*[,61] E. Zermelo, using a theorem of mechanics by H. Poincaré,[62] has shown that the usual kinetic model of a completely and permanently isolated gas behaves quasiperiodically. In other words, let us assume that the motion of the model from t_1 to t_n goes through the series of phases (11), and that in doing so H decreases from a relatively high H_1 value to a smaller value H_n. Then, after a finite[63] (although enormously long[64]) time, we will encounter in the motion left to itself a series of phases

$$\cdots (\Gamma_1), (\Gamma_2), \cdots, (\Gamma_{n-1}), (\Gamma_n), \cdots, \tag{15a}$$

all data of which sequence agree arbitrarily closely with

* "Objection or counter argument based on reasoning involving a return to the same state."

those of series (11). At the same time the motion will cover the H values

$$(16) \qquad \cdots (H_1), (H_2), \cdots, (H_{n-1}), (H_n), \cdots$$

where (H_s) is very nearly equal to H_s. It follows from $H_1 > H_n$ and from (H_1) being nearly equal to H_1 that

$$(17) \qquad (H_1) > H_n.$$

During the motion from phase Γ_n to phase (Γ_1), therefore, contrary to the formulation above of the H-theorem, the H-function has again assumed higher values.

8. Closing remarks

The monotonic decrease of the H-function (and as a consequence the irreversible approach of any distribution to the Maxwell-Boltzmann distribution) follows from the fact that in the calculations of the H-theorem the *Stosszahlansatz* is used for every successive Δt, without exception. Therefore the *Umkehreinwand* and the *Wiederkehreinwand* are mainly directed to this application of the *Stosszahlansatz.*[65]

In later developments the modified, "statistical" formulation of the H-theorem is built on completely different foundations. In these the *Stosszahlansatz* seems to have been completely eliminated.

Continuing this development, the next chapter will give the foundations (Sections 9–14) and the formulation (Sections 15 and 17) of the new approach. We will discuss only parenthetically how the *Stosszahlansatz* could be extended in the framework of this statistical formulation (Section 18).

II

The Modern Formulation of Statistico-mechanical Investigations (Kineto-statistics of the Gas Model)

9. The mechanical properties of the gas model

9a. The gas model and its phase. Let us take a gas model consisting of N identical polyatomic molecules, each having r degrees of freedom.[66] Let us define the phase (of motion) of the gas model at time t as the set of the following $2rN$ data which determine exactly the simultaneous configuration and motion of the N molecules:

(18a) $\overset{1}{q}_1, \overset{1}{q}_2, \cdots, \overset{1}{q}_r; \overset{2}{q}_1, \overset{2}{q}_2, \cdots, \overset{2}{q}_r; \cdots; \overset{N}{q}_1, \overset{N}{q}_2, \cdots, \overset{N}{q}_r$

the generalized coordinates,

(18b) $\overset{1}{p}_1, \overset{1}{p}_2, \cdots, \overset{1}{p}_r; \overset{2}{p}_1, \overset{2}{p}_2, \cdots, \overset{2}{p}_r; \cdots; \overset{N}{p}_1, \overset{N}{p}_2, \cdots, \overset{N}{p}_r$

the generalized momenta.[67]

The superscripts $1, 2, \cdots, N$ refer to the molecules while the subscripts $1, 2, \cdots, r$ refer to the degrees of freedom in the molecule.[68]

Let us further use the following notation:

(19) $\Phi = \Phi(q)$ the potential energy,

(20) $L = L(q, p)$ the kinetic energy,

(21) $E = L + \Phi$ the total energy,

of the gas model in the given phase.

The atomic motions change the configurations of the atoms ($q_1^1, \cdots, q_r^N$), while the external forces, the forces between the atoms of the same molecule and the forces which act in every collision, change the velocities (and therefore also the momenta $p_1^1, \cdots, p_r^N$). The corresponding changes in the phase of the gas model are expressed by the Hamiltonian equations of motion:[69]

$$\frac{d\overset{k}{q_s}}{dt} = \frac{\partial E}{\partial p_s^k} \qquad \frac{d\overset{k}{p_s}}{dt} = -\frac{\partial E}{\partial q_s^k}. \tag{22}$$

Unless explicitly stated to the contrary, we will assume in the following considerations that the external forces do not change with time.[70]

In this case E depends only on the q's and p's, but does not explicitly depend on t. The same holds for the quantities

$$\frac{d\overset{k}{q_s}}{dt} \quad \text{and} \quad \frac{d\overset{k}{p_s}}{dt}.$$

It is well known that the $2rN$ integrals of the equations of (22) can be expressed in the following form:[71]

$$\phi_1(q, p) \equiv E(q, p) = c_1, \tag{23a}$$

$$\begin{cases} \phi_2(q, p) = c_2 \\ \cdots\cdots\cdots\cdots\cdots, \\ \phi_{2rN-1}(q, p) = c_{2rN-1} \end{cases} \tag{23b}$$

$$\phi_{2rN}(q, p) = c_{2rN} + t. \tag{23c}$$

9b. The phase space of the gas model (Γ-*space*). Let us picture now the $2rN$ quantities (18a) and (18b), which characterize the phase of the gas model at a certain time, as the $2rN$ Cartesian coordinates of a point G in a certain Γ-space of $2rN$ dimensions.[72]

While the gas model moves according to Eq. (22), its phase point G[73] travels along a certain path[74] in Γ-space.

Its shape and position will be defined by the $2rN-1$ time-free integrals (23a) and (23b). The particular G-path lies in its entirety on the "energy surface" $E(q, p) = c_1$, and it is in fact the one-dimensional intersection of this $(2rN-1)$-dimensional hypersurface with the other $(2rN-2)$ hypersurfaces, which is given by

$$\phi_2 = c_2, \cdots, \phi_{2rN-1} = c_{2rN-1}.$$

The integral (23c) determines the time at which the G-point traverses the different points of the G-path.[75]

Let us consider now an infinitesimal $2rN$-dimensional region (A) in the Γ-space which is defined by an inequality (A) imposed on the quantities q and p. Then the volume[76] $[A]$ of the region (A) will be defined by the integral

$$(24) \qquad [A] = \int_{(A)} \cdots \int dq_1^1 \cdots dp_r^N.$$

The inequality (A) will determine on the hypersurface $E(q, p) = C_1$ a $(2rN-1)$-dimensional region $(A, E)^{-1}$. The volume $[A, E]^{-1}$ of the infinitesimal region $(A, E)^{-1}$ is defined by the integral

$$(25)^{77} \quad [A, E]^{(-1)} = \frac{Q}{\frac{\partial E}{\partial s_{2rN}}} \int_{\overline{(A, E)^{(-1)}}} \cdots \int ds_1 \cdots ds_{2rN-1}.$$

Here $s_1, \cdots, s_{2rN}$ denote the quantities $q_1^1, \cdots, p_r^N$ in any arbitrary order. The infinitesimal domain of integration, $\overline{(A, E)^{(-1)}}$ will be given by the set of values which the variables $s_1, \cdots, s_{2rN-1}$ can assume when the phase point is limited to the region $(A, E)^{(-1)}$.[78] Finally Q is used as an abbreviation for

$$(25a) \qquad Q = \sqrt{\left(\frac{\partial E}{\partial s_1}\right)^2 + \cdots + \left(\frac{\partial E}{\partial s_{2rN}}\right)^2}.$$

Because of the infinitesimal size of the region, we could take the integrand outside the integral sign.

9c. Liouville's theorem. The G-points, which at time t_0 compose the $2rN$-dimensional Γ-region (A_0) of volume $[A_0]$, occupy in the course of their motion at times $t_1, t_2, \cdots, t_s, \cdots$ the regions $(A_1), (A_2), \cdots, (A_s)$, with volumes $[A_1], [A_2], \cdots, [A_s]$, respectively. It follows from the special form of Eq. (22), which determines the flux of the G-points, that for any choice[79] of the (A_0) we have

$$[A_0] = [A_1] = \cdots = [A_s]. \tag{26}$$

This is Liouville's theorem. It can be stated as follows:[80] The streaming of the G-points in the Γ-space as given by Eq. (22) generates a continuous point transformation, which transforms each $2rN$-dimensional region into another one of the same volume.[81]

Similarly, if one considers the G-points which at time t_0 occupy an infinitesimal, $(2rN-1)$-dimensional region $(A, E)_0^{(-1)}$ of volume $[A, E]_0^{(-1)}$, one obtains the following immediate consequence of Eq. (26):[82]

$$\frac{[A, E]_0^{(-1)}}{Q_0} = \frac{[A, E]_1^{(-1)}}{Q_1} = \cdots = \frac{[A, E]_s^{(-1)}}{Q_s}. \tag{27}$$

9d. Stationary density distributions in the Γ-space.[83] Let us assume that at time t_0 we distribute an infinite number of G-points over the unlimited Γ-space in such a way that by assigning a suitable measure we can define a certain "spatial" density ρ_0 for each point in the Γ-space.[84] If we select the function $\rho_0(q, p)$ quite arbitrarily at time t_0, then in general in the subsequent streaming of the G-points, the density in each point of Γ-space will change in time.[85]

It follows from Eq. (26) that

$$\frac{d\rho}{dt} = 0. \tag{26'}$$

The distribution of the "spatial" density ρ will be stationary with respect to the streaming described by Eq. (22) if, and only if, $\rho_0(q, p)$ is chosen to be constant along each single G-path. (For different G-paths we may arbitrarily choose different values of $\rho_0(q, p)$.) This means that $\rho_0(q, p)$ should have the form

$$\rho_0(q, p) = F(E, \phi_2, \cdots, \phi_{2rN-1}) \tag{28}$$

where F is an arbitrary, single-valued function of the $2rN-1$ variables.[86]

Similarly, one can use the "surface density" $\sigma_0(q, p)$ to describe the distribution of an infinite set of G-points on the $(2rN-1)$-dimensional hypersurface $E(q, p)=c_1$. The necessary and sufficient condition for the stationary behavior of such a "surface density" σ is given by[87]

$$\sigma_0(q, p) = \frac{1}{Q(q, p)} F(\phi_2, \cdots, \phi_{2rN-1}). \tag{29}$$

Here F is an arbitrary, single-valued function of the $(2rN-2)$ variables; Q is defined by (25a); (q, p) is restricted by the condition $E(q, p)=c_1$.

10. The gas model as an ergodic system

10a. Ergodic mechanical systems.[88] One can define mechanical systems by the following property: The G-path describing their motion is an open curve[89] that covers every part of a multidimensional region densely.[90] Prompted by the existence of such systems, Boltzmann[91] and Maxwell[92] defined a class of mechanical systems in the following way:

The single, undisturbed motion of the system, if pursued without limit in time, will finally traverse "every phase point" which is compatible with its given total energy. A mechanical system satisfying this condition is called by Boltzmann an "ergodic system."[93]

Boltzmann and Maxwell infer from this definition the following corollaries:

1. For an ergodic system all motions with the same total energy take place on the same G-path.[94]

2. This means that all these motions differ only in the value of the constant c_{2rN}, which appears as a constant additive to the time (cf. Integral 23c).[95]

3. All of these motions yield the same value for the time average of any function $\phi(q, p)$ of the phase variables.[96]

It is on account of this last property that the definition of ergodic systems and the assumption that the gas models are ergodic appear in Boltzmann's investigations (cf. Section 11).

However, the existence of ergodic systems (i.e., the consistency of their definition) is doubtful. So far, not even one example is known of a mechanical system for which the single G-path approaches arbitrarily closely each point of the corresponding energy surface.[97] Moreover, no example is known where the single G-path actually traverses all points of the corresponding energy surface.[98] Nevertheless, not only is the latter the wording of the Boltzmann-Maxwell definition, but precisely this feature of the definition serves as the basis for the statement of the two authors that in the gas model, as an ergodic system, all motions with the same total energy traverse the same ϕ-path and hence give the same time average for every $\phi(q, p)$.[99]

10b. Ergodic density distributions in Γ*-space.* The following special cases of stationary density distribution have been given chief consideration in the literature:

$$(30) \qquad \text{(cf. 28)} \qquad \rho(q, p) = F(E)$$

for spatial density of a distribution in Γ-space,[100] and

$$(31) \qquad \text{(cf. 29)} \qquad \sigma(q, p) = \frac{1}{Q(q, p)}$$

for the surface density of a distribution over the energy surface

$$E(q, p) = E_0.$$[101]

We want to make the following remarks in connection with this special choice of density distributions:

1. It is decisive in the development of the main results in the statistical theory for the motion of gases. (Cf. note 172.)

2. When Maxwell and Boltzmann first introduced these special density distributions, they justified them by referring explicitly to the hypothesis that the gas models are ergodic systems.[102]

This special choice is generally taken over in the later literature, even though the ergodic hypothesis is then not even mentioned.[103]

It is advisable therefore to call the density distributions (30) and (31) "ergodic" in order to remind ourselves that so far we have no other justification for their choice than the invocation of the ergodic hypothesis.

11. The average behavior of the gas model for a motion of infinite duration

11a. Boltzmann's investigations. Boltzmann began studying this problem even before the establishment of the H-theorem, when he tried to show that only the Maxwell-Boltzmann distribution corresponds to thermal equilibrium.[104]

The starting point of this investigation may have been the following statements of a generalized empirical fact:[105] An isolated gas quantum which is in a state different from thermal equilibrium will go over into thermal equilibrium and will permanently stay there. If we consider therefore the average behavior of a gas quantum that has been left to itself for a long period of time T, we find that if T goes to infinity the average behavior will be identical with the behavior in thermal equilibrium.

Therefore the aim of the above-mentioned investigations by Boltzmann seems to be to obtain proof of the following statement:

The average behavior of a gas model during a motion of unlimited duration corresponds to the Maxwell-Boltzmann distribution of state.

The fundamental assumption underlying this investigation is the hypothesis that the gas models are ergodic systems (cf. Section 10). With the help of this hypothesis Boltzmann computed the time average of, for instance, the kinetic energy of each atom (the same value is obtained for all atoms!).[106] Likewise he calculated the time average of other functions $\phi(q, p)$ which characterize the average distribution of state.

Boltzmann, in order to calculate this time average, uses the picture of a set of an infinite number of identical samples of the gas model in question. These samples move, completely independently of each other,[107] in such a way that they all have the same total energy E_0, and that their G-points are distributed at time t_0 over the surface $E(q, p) = E_0$ with the "surface density"

$$\text{(31a)} \qquad \text{(cf. 31)} \qquad \sigma = \frac{1}{Q(q, p)} \cdot$$

According to Section 9d this distribution is a stationary one.[108]

For such a phase distribution of the set of systems we can obtain the "ensemble average" (i.e., the average over the set of systems) of the above-mentioned phase functions by integrating over the energy surface:[109]

$$\text{(32)} \qquad \overline{\psi(q, p)} = \frac{\int \psi \cdot \sigma \cdot ds}{\int \sigma ds} \cdot$$

In order to get from this ensemble average to the time average, which is what Boltzmann wants, one needs the following chain of equalities:

(33) Ensemble average = the time average of the ensemble average
= the ensemble average of the time average
= time average.

The first equality follows from the stationary character of the phase distribution,[110] the second because the forming of averages is commutative. The third equality, however, is based on the statement that all motions of the set in question give the same time average for $\psi(q, p)$.

It is at this point that the hypothesis about the gas model being ergodic enters. Because of the doubts about the internal consistency of the ergodic hypothesis, this investigation cannot be considered free of objection (Section 10a).[111]

Incidentally, the outlined way of calculating the time average is closer to the arrangement which can be found in the work of Maxwell (1879),[112] and in the subsequent treatments by Lord Rayleigh[113] and by Jeans.[114] However, they differ from Boltzmann's original treatment only in a formal way. Using the picture of an ergodically distributed ensemble of systems and the hypothesis of the gas model as an ergodic system,[115] Boltzmann derives first a formula for the time intervals which the G-point during its motion of unlimited duration spends in the various parts of the energy surface:[116]

$$\lim_{T=\infty} \frac{dt}{T} = \frac{\sigma ds}{\int \sigma ds} \tag{34}$$

where σ is given again by Eq. (31a). For the time average of a phase function $\psi(q, p)$ this formula obviously gives again Eq. (32).

11b. Criticism and meaning of Boltzmann's results. The following remarks are in order concerning Eq. (34):

1. If one accepts the ergodic hypothesis, then Eq. (34) becomes a purely mechanical theorem independent of any "probability" considerations.

2. If one discards the ergodic hypothesis, or tries to retain it in a modified form,[117] then for the time being one lacks any criterion to decide whether Eq. (34) is still valid or even whether it represents a somehow useful approximation.

If we accept the validity of Eq. (34), we can conclude from it more than just a statement about average behavior. In fact, it also determines essentially the relative time intervals that the gas spends in the various distributions of state.

This deeper formulation of the question was shelved in Boltzmann's investigations by the soon emerging formulation of the H-theorem (1872). Boltzmann took up this problem again only to counter Loschmidt's *Umkehreinwand* (Section 7b) and to obtain a modified formulation of the H-theorem. He tried to show that, if we consider a motion of unlimited duration, then the Maxwell-Boltzmann distribution very strongly dominates in time over all other distributions, and hence the tendency to approach this particular distribution is quite understandable.

We will discuss all this in detail in Sections 13 and 14; before doing so, however, we want to discuss certain problems which will be of help later.

12. *Mechanical properties of the gas model* (continued)

12a. The phase space of the molecule (μ-space): The state distribution **Z** *of the molecule.* A point in a $2rN$-dimensional "Γ-space" corresponded to the phase of the whole gas model. It is now useful to describe the instantaneous phase of each of the N molecules (e.g., the phase of the kth by a point $m^{(k)}$) in a $2r$-dimensional "μ-space." We will assign to each molecule an image point $m^{(k)}$ which is permanently connected with the molecule through the index k. In order to determine the phase of the whole gas we have to specify the position of all of the N points $m^{(1)}, \cdots, m^{(N)}$.[118]

Let us divide the μ-space once and for all into very small but finite[119] and identical parallelopipeds ω.[120] Each cell ω_i contains at time t a certain number a_i of image points $m^{(k)}$.[121]

The state distribution **Z**[122] of molecules at time t is defined as the set of numbers a_i.

12b. The volume in Γ*-space corresponding to a state distribution* **Z**.[123] It follows from the definition of the distribution of state that—

1. To a given Γ-point corresponds a uniquely determined state distribution **Z**.

2. To a given state distribution **Z** corresponds a $2rN$-dimensional continuum of Γ-points, the domain (**Z**), which one may call a "**Z**-star."[124]

3. The volume [**Z**] of the **Z**-star is given by[125]

$$[\mathbf{Z}] = \frac{N!}{a_1!a_2! \cdots} [\omega]^N. \tag{35}$$

12c. Functions of the state distribution. If we displace the phase point G continuously every time the G-point passes from one **Z**-star to another, the set of integers a_i changes in a discontinuous way. Hence every function of the numbers a_i—or function $F(\mathbf{Z})$ of the distribution of state—if considered as a function of the phase (q, p), will be a discontinuous phase function.

These functions are characteristic tools of the kinetic theory.

The value of $[\mathbf{Z}]$ or rather its first factor,

$$P(\mathbf{Z}) = \frac{N!}{a_1!a_2! \cdots}, \tag{36}$$

serves as an example for a $F(\mathbf{Z})$.

All the $F(\mathbf{Z})$'s which one uses in the kinetic theory of gases to approximate the exact values of continuous phase functions are also in this category. Let us denote by λ_i the kinetic energy of a molecule when its phase point is exactly at the center of cell ω_i in the μ-space. Since very small ω cells are chosen, the total kinetic energy of all a_i molecules whose phase point lies in the cell ω_i can be approximated very well by

$$a_i\lambda_i. \tag{37}$$

Hence the exact value of the kinetic energy of the whole gas model, $L(q, p)$—a continuous phase function—can be approximated by the $F(\mathbf{Z})$'s:

$$L(\mathbf{Z}) = \sum a_i\lambda_i. \tag{38}$$

Similarly the potential energy $\Phi(q, p)$ can be approximated by

$$\Phi(\mathbf{Z}) = \sum a_i\phi_i, \tag{39}$$

and hence the total energy $E(q, p)$ by

$$E(\mathbf{Z}) = \sum a_i\epsilon_i. \tag{40}$$

One can then easily verify the following statement, which will be used in Section 13:

The equation

$$E(\mathbf{Z}) = E_0 \tag{41}$$

cuts a $2rN$-dimensional (shell-like) domain out of the Γ-space; this domain is the union of all $\mathbf{Z}$-stars for which the a_i's satisfy Eq. (41). On the other hand, the equation

$$E(q, p) = E_0 \tag{41a}$$

defines a $(2rN-1)$-dimensional hypersurface, which is partly inside, partly outside, the shell given by Eq. (41).

12d. The function $H(\mathbf{Z})$. Equations (35) and (36) depend on the instantaneous state distribution $\mathbf{Z}$ only through the function

$$\prod(\mathbf{Z}) = a_i!a_2!a_3! \cdots.$$

If, therefore, we use the Stirling approximation for the logarithm of $\prod(\mathbf{Z})$:

$$\text{(42)} \quad \log \prod(\mathbf{Z}) = \sum \left\{\tfrac{1}{2}\log 2\pi + (a_i + \tfrac{1}{2})\log a_i - a_i\right\}$$

and we collect in this all the terms which are decisive in determining the variation of $\log \prod(\mathbf{Z})$ with $(\mathbf{Z})$, then we get the function

$$\text{(43)} \qquad H(\mathbf{Z}) = \sum_i a_i \log a_i.$$

The works of Boltzmann have revealed the widespread significance of this function.[126]

12e. The symbols $dH(\mathbf{Z})/dt$ *and* $\Delta H(\mathbf{Z})/\Delta t$. As it will be shown in Section 14b, the interchange of a certain time difference quotient of $H(\mathbf{Z})$ with a differential quotient has caused a considerable amount of confusion in the discussion of the H-theorem. One can avoid such a confusion by making the following remarks:

1. The curve describing the time behavior of a certain $F(\mathbf{Z})$ is always a step function.[127] The time differential quotient therefore can assume only the following three values:

$$\text{(44)} \quad \frac{dF(\mathbf{Z})}{dt} = 0, \quad \frac{dF(\mathbf{Z})}{dt} = -\infty, \quad \frac{dF(\mathbf{Z})}{dt} = +\infty.$$

2. The "time element Δt" used in the kinetic theory of gases, and in particular in the H-theorem, although very

small compared to the time intervals used experimentally, cannot be made arbitrarily small.[128]

3. The statements about the time behavior of $H(\mathbf{Z})$ or of an arbitrary $F(\mathbf{Z})$ usually therefore refer to a discrete sequence of points selected from the step function, and one has to consider difference quotients whose Δt still contains a very large number of steps of the step function.[129]

13. The dominance of the Maxwell-Boltzmann distribution

As a by-product of the attempt to prove the statement introduced at the end of Section 11b, Boltzmann (1877)[130] discovered the following property of the Maxwell-Boltzmann distribution:

(I) Of all the state distributions $\mathbf{Z}$, which have a certain value

$$E(\mathbf{Z}) \equiv \sum a_i \epsilon_i = E_0, \tag{45}$$

the Maxwell-Boltzmann distribution

$$a_i = Ae^{-h\epsilon_i} \tag{46}$$

has the largest value of $[\mathbf{Z}]$. The value of $[\mathbf{Z}]$ decreases very rapidly if the a_i's, holding $E(\mathbf{Z})$ constant, differ appreciably from the Maxwell-Boltzmann values (Eq. 46).[131]

In the usual terminology, which does not take into account the difference between the two equations $E(q,p) = E_0$ and $E(\mathbf{Z}) = E_0$ (cf. the end of Section 12c), statement (I) above can hardly be distinguished from the following statement:

(II) On any hypersurface $E(q, p) = \text{constant}$, the "surface" area[132] of the regions corresponding to the Maxwell-Boltzmann distribution will be enormously larger than the area of all other regions.[133]

If we accept the dubious ergodic hypothesis and Eq. (34) (Section 10), then from statement (II) follows immediately the result announced at the end of Section 11b.

(III) If we consider a motion of the gas model which is of unlimited duration, the Maxwell-Boltzmann distribution will predominate overwhelmingly in time over all other appreciably different state distributions.

If, however, we discard the use of the ergodic hypothesis, as the newer works seem to do, then there is a definite gap between (II) and (III).

14. The modified formulation of the H*-theorem*

In the usual probabilistic terminology this gap between statements (II) and (III) is hardly noticeable, because the term "relative probability" is used in a variable meaning. First it is an abbreviation for the ratio of certain Γ-volumes, as in the following formulation of Theorem (I):

(I′) Of all state distributions having the same total energy, the Maxwell-Boltzmann distribution is the "most probable" one, in fact overwhelmingly the most probable one.

Subsequently, however, this "probability" is interpreted to mean either the ratio of time intervals or the relative frequencies in other, very different statistical ensembles. In this way formulation (I′) leads to statement (III) and also to all other further statements which, together, make up the modified formulation of the *H*-theorem.

From a logical viewpoint such a procedure is not very satisfactory. Therefore the terminology using probabilities will be eliminated in the subsequent discussions. In this way the modified version of the *H*-theorem will appear as a sequence of hypothetical statements about a certain group of motions of the gas model (statements IV–VII). Only by this method can we survey the gaps in the new formulation.

14a. The step function of the H(**Z**) *values.* Let us use the quantity $H(\mathbf{Z})$ as the measure for the deviation from the Maxwell-Boltzmann distribution of a state distribution (**Z**). Then statement (III) assumes the following form:

The step function[134] which describes the time dependence of $H(\mathbf{Z})$ for an indefinitely continuing motion of the gas model,

(IVa) Will remain during overwhelmingly long time intervals very close to the minimum value H_0. Only during relatively short periods will it rise appreciably or even strongly from the minimum value.

(IVb) Let us denote by H_1 a value which is considerably different from the minimum value H_0. The sum of the time intervals during which the step function is above the horizontal line H_1 will decrease very rapidly if we raise the height of the H_1 line by even a very small amount.

14b. The H*-curves.* Let us now select a discrete set of points on this step function, with an equal distance Δt between the abscissae of two neighboring points, where Δt is chosen in accordance with Section 12e. We can then repeat statement (IV) for these points, but in a more precise way. Let us denote three neighboring H-values by

$$H_a < H_b < H_c,$$

and let us assume that all three of these are quite different from the minimum H_0. Then, following Boltzmann,[135] we can formulate statement (IV) in the following way:

If we consider in this discrete set of points all those points that lie at a height H_b, then we see that in an overwhelming number of cases[136] they form maxima, schematically represented by

(47)
$$\begin{array}{ccc} & H_b & \\ H_a & & H_a. \end{array}$$

Only a very small fraction of the points lies on a decreasing slope.

(48)
$$\begin{array}{ccc} H_c & & \\ & H_b & \\ & & H_a \end{array}$$

or on an increasing slope

(49)
$$\begin{array}{ccc} & & H_c, \\ & H_b & \\ H_a & & \end{array}$$

and an even smaller fraction forms a minimum

(50)
$$\begin{array}{ccc} H_c & & H_c. \\ & H_b & \end{array}$$

From now on we will denote this and only this discrete set of points selected from the $H(\mathbf{Z})$ step function by the term "*H*-curve."

One can then express the above statements in still another way which is the way Boltzmann generally used:

(Va) The "*H*-curve" almost always decreases immediately from each point H_1 which lies above the minimum H_0.[137]

(Vb) This statement is equally valid regardless of whether one traverses the curve from left to right (positive time sequence) or from right to left (inverted time sequence).

(Vc) Otherwise the *H*-curve, just as the step function, almost always runs very near the minimum H_0.

Statements (Va) and (Vb) were occasionally declared to be geometrically meaningless. In this connection let us make the following remarks:

1. Boltzmann used the term "H-curve" in a variable meaning—
   a) for the step function,
   b) for the discrete set of points selected thereof, and
   c) for a certain "smoothed" interpolating curve of a) and b),[138]

and he also neglected sometimes to make a clear distinction between the "H-curve" and the "curve of the H-theorem" (cf. Section 14d).

2. Boltzmann's language is very apt to create the completely false impression that the "H-curve" appearing in (Va) and (Vb) is the smooth interpolating curve.

3. In particular, he promoted this misunderstanding by always calling these maxima of the H-curves "humps," which makes one think almost necessarily of a maximum with a horizontal tangent.

If, however, one means by the H-curve exclusively and consistently a discrete set of points and then realizes that all of its higher-lying points form almost exclusively point maxima, then all the objections against the geometrical admissibility of statements (Va) and (Vb) disappear.

This point is essential for the understanding of Boltzmann's position with respect to the *Umkehreinwand* and the *Wiederkehreinwand.*[139]

14c. The bundle of H*-curves: Its concentration curve.* Let us consider a distribution $\mathbf{Z}_A$ of state with a relatively high $H(\mathbf{Z}_A)$ at t_A. Let us find the corresponding $\mathbf{Z}_A$-star in the Γ-space (cf. Section 12b). The continuum of the motions of all the points in the star produces a bundle of H-curves in the (t, H) plane which radiate out of a point $t=t_A, H=H(\mathbf{Z})_A$. We make the following assertions about its behavior:[140]

(VIa) The totality of H-values to which the bundle leads at later times $t_A+n\Delta t$ converges with an extremely small dispersion around a certain value $\mathcal{H}_n$.[141]

Let us collect these values $\mathcal{H}_1$, $\mathcal{H}_2$, $\mathcal{H}_3$, $\cdots$ into a discrete set of points and call this briefly the concentration curve of the bundle. Then we assert in addition:

(VIb) The concentration curve of the bundles monotonically decreases from the high initial value $H(\mathbf{Z}_A)$, converges to the minimum H_0, and never again departs from it.[142]

Furthermore, we can elaborate on (VIa):

(VIc) From t_A on, over a long time, an overwhelming majority[143] of the curves will run very near to the concentration curve.

(VId) Those motions, on the other hand, whose H-curves run without interruption very near to the concentration curve even if the motion is followed over an unlimited time interval, if they exist at all, form a set with a measure of a lower order.[144]

14d. The curve of the H*-theorem.* In the context of the formulation which we have used since Section 9, the *Stosszahlansatz* (Section 3b), and with it the whole H-theorem, are still a meaningless computational scheme.[145] The initial distribution $\mathbf{Z}_A$ defines a new $\mathbf{Z}$ for the end of each successive interval Δt. Let us denote these distributions by $\mathbf{Z}_1$, $\mathbf{Z}_2$, $\cdots$. One thus obtains in addition to the previous H-curves a new H-sequence, which is discrete (the spacing of the abscissa$=\Delta t$) and which decreases monotonically (cf. Section 6). We call this the "curve of the H-theorem."

This generates a statistical formulation of the H-theorem through the following assertion, which is again unproved:

(VII) The curve of the H-theorem is identical with the concentration curve of the bundle of H-curves defined by (VI).

15. The statistical character of kinetic interpretations

The kinetic interpretation of an aerodynamic process, just as any other "explanation" of a physical phenomenon, consists of the representation of the observed sequence of states by a purely conceptual scheme. A special feature of kinetic interpretations, however, is the statistical character of these schemes. Each single process which takes place in the given gas quantum is made to correspond to a whole group of motions of the gas model. This is done with the help of the following concluding assertion:

(VIII) The actually observed sequence of states of a gas quantum from time t_A on is identical with that created by the overwhelming majority[146] of the motions discussed in (VI).

15a. Distribution of state and observable data. The representation of the aerodynamic processes formulated in (VIII) reveals the following gap: Assertions (V)–(VII) deal with a continuum of motions which are generated by a distribution of state $\mathbf{Z}_A$ prescribed for t_A. The actual observations, on the other hand, do not give a distribution of state but determine what might be called the "visible" state of the gas quantum. It is a rough determination of the distribution of pressure, density, temperature, and velocity of flow inside the gas.

In the usual treatments this gap is overlooked in the further developments. The procedure is carried out as if the following theorem could be proved:

(IX) Of all the various distributions of state $\mathbf{Z}$ which correspond to a certain "visible" state S_n, there is a special one, $\mathfrak{Z}_n$, such that a very much larger region in Γ-space belongs to this $\mathfrak{Z}_n$ and to the $\mathbf{Z}$'s very near to it than to all other $\mathbf{Z}$'s belonging to S_n.

15b. Postulate of determinacy: The Brownian motion. One

can abstract from experience the following statement about the behavior of an isolated gas quantum:

A "visible" state S_A at time t_A completely determines the "visible" state for any subsequent time t_B.

This statement is clearly not a direct expression of an experimental fact. Apart from a very small group of aerodynamical processes, we always deal with turbulent motions of the gas, where it is impossible to follow the "visible" state with our measuring instruments. Furthermore, even the best insulation against thermal conduction and radiation is completely unsatisfactory unless we deal with very short periods of observation.

The above statement is therefore a postulate, which goes considerably beyond the possibility of an experimental check. Our observations cannot tell us anything about what the sequence of state of an actual isolated gas quantum would be over a very long period of time and whether it would satisfy the principle of determinacy.

As soon as we include a microscope among the instruments of observation of the "visible" state, we discover the Brownian motion. For such a more precise "visible" state the principle of determinacy does not seem to hold any more. On the other hand, it has been shown that this phenomenon is surprisingly well suited to a statistical interpretation.[147]

16. A retrospective view of the Umkehreinwand *and* Wiederkehreinwand

Do these objections show that there are inner inconsistencies in the statistical interpretation? Several respected investigators answer this question in the affirmative even today.[148] A careful consideration of the geometrical properties of the H-curves (as given above) is needed in order to understand how Boltzmann in his elucidations of the H-theorem could assign to these H-

curves the following apparently self-contradictory properties:

1. That H almost always immediately decreases from any higher value (in the sense of the H-theorem).

2. That this holds both in the direct and reverse time directions (as Loschmidt demands it—Section 7a).

3. That H is quasiperiodic (as Zermelo demands it—Section 7b).

As soon as we clear up all the misunderstandings in this respect (see the discussion of Section 14b), we have to agree with Boltzmann that we can find no inner inconsistencies this way at all.[149]

There is another formulation of the *Umkehreinwand* which has to be discussed.[150] Let us consider all those phase points in Γ-space which correspond to a given high H-value H_A. Let us assume that for each of these points the change ΔH for the following time interval Δt is determined, and let us form the average of this over the set of phases in question. Supposedly it can be shown that this average value is zero instead of being smaller than zero.

It is true that to each phase Γ in the set of phases in question we can find a phase Γ′ in such a way that Γ and Γ′ assign the gas model the same configurations but opposite velocities.[151] However, the statement that

$$\left(\frac{\Delta H}{\Delta t}\right)_{\Gamma'} = -\left(\frac{\Delta H}{\Delta t}\right)_{\Gamma} \tag{51}$$

is incorrect.[152] Therefore the consequence, that the average value is zero, is also untrue.

Although these objections do not prove the existence of internal inconsistencies, they emphasize the fact that consistency is assured only by proof of assertions (III)–(VII).

17. The relationship of the statistical interpretation to the entropy theorem

If we restrict ourselves to time intervals which are experimentally realizable, the statistical interpretation is equivalent to the requirement that the entropy of an isolated gas quantum should always increase.[153] Beyond that, however, one can in fact defend two opposite points of view.

1. One can assert that, if we observed a perfectly isolated gas quantum over an unlimited period of time, it would behave in a quasiperiodic fashion with very long periods.[154] This point of view has the advantage of clearly revealing an essential feature of the theory as formulated above. On the other hand, it reminds us to what a large extent the postulate about the entropy increasing without exception is an extrapolation from the experiments.[155]

2. One can discard the claim that the relatively primitive assumptions about the structure of the gas model also give a correct picture of the phenomena even over very long time intervals. This point of view was, of course, also considered by Boltzmann. He emphasized very early (1871)[156] that in the further development of the kinetic theory one has to consider the interaction of the molecules and the ether (i.e., the influence of radiation on the thermal equilibrium). However, in the discussions about the H-theorem, he was right to insist on the first point of view to its final consequences. In this case a reference to, for instance, thermal radiation would easily lead to a premature condemnation of Boltzmann's ideas, as if the increase in entropy for processes during an observable time interval could not be interpreted without invoking radiation.

18. *Further statistical development of the* Stosszahlansatz: *Hypothesis of molecular chaos*

18a. Boltzmann's ideas. Boltzmann repeatedly touched upon the question of how the *Stosszahlansatz*, fundamental to the older formulation of the H-theorem, should be reinterpreted in the light of the new, statistical point of view. The objections directed against the H-theorem have certainly shown that the free motion of the gas model cannot always obey the *Stosszahlansatz*.[157] Using the terminology of probability, one can, perhaps, summarize Boltzmann's ideas in the following way:[158]

1. The *Stosszahlansatz* gives for each time interval Δt only the "most probable" value of the number of collisions. (Correspondingly the H-theorem gives for each Δt only the most probable value of the change in H.)

2. The actual number of collisions (and the actual change in H) fluctuates about this most probable value and can also assume other values with a small, but nonzero probability.[159]

The relative probabilities of the various changes in H should at the same time be in agreement with the relative probabilities of the different values of H as they are given by statement (I′) of Section 14.

However, here again large gaps appear as soon as we replace, in the above statements, the abbreviating term "probability" by giving the corresponding "frequencies."

In this sense Jeans[160] has made statement (1) above more precise. Statement (2), on the other hand, which, in our eyes, represents what Boltzmann actually meant by the "hypothesis of molecular chaos,"[161] is still awaiting a corresponding formulation. The following considerations are based mainly on the work of Jeans and attempt to establish a connection with the criticisms which Burbury[162] has repeatedly made of the *Stosszahlansatz*.

18b. More precise determination of the distribution of state: The Jeans grouping. The knowledge of the distribution of state $\mathbf{Z}_A$ is insufficient to determine the number of collisions for the subsequent time interval Δt. For that we also have to know how many pairs of molecules are in the process of making collisions of the various kinds at time t_A.[163] We might abbreviate this as the "grouping" of the molecules.

The $\mathbf{Z}_A$-star which in the Γ-space belongs to the distribution of state $\mathbf{Z}_A$ (cf. Section 12b) can be divided into subregions corresponding to the various groupings. Of these subregions let us consider the one occupying the largest Γ-volume. From now on we will call this the "Jeans grouping." Then the following assertions can be made.[164]

(Xa) Almost the whole volume of every $\mathbf{Z}_A$-star is taken up by the corresponding Jeans grouping and those very close to it. Only a relatively very small part is occupied by groupings appreciably different from the Jeans grouping.

(Xb) The Jeans grouping gives for the subsequent time interval Δt exactly that system of collision numbers which satisfies the *Stosszahlansatz.*[165]

18c. The hypothesis of molecular chaos. Let us now return to the group of motions we discussed in Section 14c.

At initial time t_A this group of motions originates from all those phases which correspond to a given distribution of state $\mathbf{Z}_A$, i.e., from that exact set to which statements (X) apply. It follows from (X) therefore that in the first time interval, i.e., from t_A, to $t_A + \Delta t$, the overwhelming majority of the motions under consideration will satisfy the *Stosszahlansatz.*

At a later time t_B the paths of the group under consideration will have reached various Γ-points corresponding to distributions of state $\mathbf{Z}_{B'}$, $\mathbf{Z}_{B''}$, etc., which are already different from each other. Conversely, at this moment each single distribution of state of the group, e.g., $\mathbf{Z}_{B'}$, will be realized only by a part of the phases that compose the corresponding **Z**-star.

We therefore need a supplementary statement, which can be given as follows:[166]

(XI) In each of these subsets the various groupings occur with the same relative frequency as in the total set to which statements (X) apply.

Only if we also accept statement (XI) can we justify the assertion that the overwhelming majority of motions under consideration satisfy the *Stosszahlansatz* not only in the first time interval from t_A to $t_A+\Delta t$, but also in the subsequent ones.

Statement (XI) may need some quantitative corrections if some remarks made by Burbury are taken into account.[167] According to Burbury, a "correlation of velocities" is created by the finite dimension of the molecules and by the forces with which they act on each other. This correlation influences the number of collisions. Since Jeans's analysis does not claim to go beyond the proof of (Xa) and (Xb), it cannot give any information about these corrections.

Although the formulation of the "hypothesis of molecular chaos" still contains many gaps, it certainly shows clearly that the *Umkehreinwand* and the *Wiederkehreinwand* affect only the original formulation of the *Stosszahlansatz*; in fact, they prove its untenability. The improved statistical version of the *Stosszahlansatz*, however, takes into account all those requirements arising from these objections.[168]

III

W. Gibbs's *Elementary Principles in Statistical Mechanics*

19. The problem of axiomatization in kineto-statistics

In order to evaluate properly the transition from the older formulation of kinetic theory to the still sketchy statistical formulation we need a comparison with other branches of science in which the methods of probability theory have found an application. All these show a similar process of development.

At the beginning the term "probability" appeals explicitly to a certain feeling of estimation which is expected to be able to fill in gaps in the observations and calculations.[169] Above all, it appeals to a certain instinctive knowledge to the effect that elementary occurrences should in every instance be "equally possible."[170] Later there is a critical reaction which leads in the various domains of application to very different results. However, very seldom does this reaction lead to a total rejection (this seems to be the case, e.g., for the "Theory of Decisions by Jury" and the "Theory of the Statements by Witnesses").[171] Everywhere that the former assumptions had been proved by experience to be fruitful, a new formulation was found in which the contested assumptions were maintained and further developed (see the formulations which in mathematical statistics (*Kollektivmasslehre*) are introduced for the description of various mass phenomena).[172]

Only a sketch of the corresponding further development of the kinetic theory is at present available, and it is doubtful how much of it can be carried through. In any case, it indicates clearly the form that the former "probability hypotheses" and the "hypotheses" about the nature of the gas molecule (cf. Section 1) will have to assume.

The kinetic "explanations" become representations or mappings of some conceptual scheme (cf. Section 15), and correspondingly the two groups of hypotheses become more or less arbitrary assertions about the structure of this conceptual scheme. These assertions will be—

1. About the structure of the gas model.
2. About the selection of the group of motions.[173]

Freedom in the choice of these assertions seems to be restricted essentially by only one requirement: the scheme has to be self-consistent. This tendency to axiomatize is an important factor throughout the new development of the kinetic theory. It first attracted the general attention of mathematicians[174] after the appearance of the program formulated by W. Gibbs in the preface of his *Elementary Principles of Statistical Mechanics* (1901).

20. W. Gibbs's program in his Statistical Mechanics

In his preface Gibbs describes the purpose of his treatise somewhat as follows: The statistico-mechanical concepts and methods have so far been developed not as an independent system but only as an aid for the kinetic theory of matter. In this manner of developing the theory grave difficulties arose from the attempt to establish hypotheses about the structure of the gas models in such a way that they would account, if possible, for all experimental results.

Difficulties of this kind have deterred the author from attempting to explain the mysteries of nature, and have forced him to be contented with the more modest aim of deducing some of the more obvious propositions relating to the statistical branch of mechanics. Here, there can be no mistake in regard to the agreement of the hypotheses with the facts of nature, for nothing is assumed in that respect. The only error into which one can fall, is the want of agreement between the premises and the conclusions, and this, with care, one may hope, in the main, to avoid.*[175]

We will restrict the subsequent discussion of the contents of Gibbs's book to that necessary for the critical study of the following two questions:

1. To what extent has Gibbs achieved his announced goal of founding a statistical mechanics free of internal contradictions?

2. What is the relationship between the analogies to thermodynamics given by Gibbs and those given by Boltzmann?

Especially with regard to the first question, we have to distinguish three groups of investigations which are combined in Gibbs's discussions:

a) The introduction by definition of certain special stationary ensembles of systems (stationary density distributions in Γ-space) and the development of theorems about the various average values which can be derived for these ensembles because of their special nature.

b) Statements about the behavior of nonstationary ensembles; in particular, statements about how a nonstationary density distribution in Γ-space is gradually disarranged and consequently how certain functions of this density distribution change in time.

c) Considerations which are meant to show the existence of a deep analogy between the behavior of those special ensembles and the thermodynamic behavior of macroscopic systems.

* In the original German text this passage is quoted in Zermelo's translation. I have quoted the original words of Gibbs (Pref., p. x). —THE TRANSLATOR

21. The introduction of certain special stationary density distributions in Γ-space (canonical and microcanonical distributions)

Gibbs, after recapitulating Boltzmann's investigations of the most general stationary density distribution in Γ-space

$$\rho(q, p) = F(E, \phi_2, \cdots, \phi_{2rN-1}), \tag{52}$$

restricts himself essentially to the following very special distributions:[176]

A) The microcanonical distribution: ρ is everywhere zero except between the two energy surfaces $E=E_0$ and $E=E_0+\delta E_0$, where δE_0 is very small. Within this shell ρ has a constant value. This distribution of volume density $\rho(q, p)$ for $\delta E_0=0$ is obviously equivalent to the ergodic surface distribution (Eq. 31) of Section 10b.

B) The canonical distribution:[177]

$$\rho(q, p) = e^{(\Psi-E)/\Theta}. \tag{53}$$

Here $E(q, p)$ is the total energy the gas model has if its phase point G in Γ-space lies at (q, p); Θ is an arbitrary constant, the "modulus" of the distribution (Eq. 53); and Ψ is a second constant, which is determined from the requirement that the integral of

$$\rho \cdot d\overset{1}{q_1} \cdots d\overset{N}{p_r},$$

if extended over the whole Γ-space, should be

$$\int \cdots \int \rho d\overset{1}{q_1} \cdots d\overset{N}{p_r} = 1 \tag{54}$$

i.e., from

$$\int_{-\infty}^{+\infty} \cdots \int_{-\infty}^{+\infty} e^{(\Psi-E)/\Theta} d\overset{1}{q_1} \cdots d\overset{N}{p_r} = 1. \tag{55}$$

By this arrangement we have simplified the calculation

of the average value of a phase function $f(q, p)$ over the ensemble

$$(56) \qquad \overline{f(q,p)} = \frac{\int_{-\infty}^{+\infty} \cdots \int f \cdot e^{(\Psi - E)/\Theta} \, d\overset{1}{q_1} \cdots d\overset{N}{p_r}}{\int_{-\infty}^{+\infty} \cdots \int e^{(\Psi - E)/\Theta} \, d\overset{1}{q_1} \cdots d\overset{N}{p_r}}$$

by reducing the denominator to the convenient value of 1.

With an eye on later applications we will assume that the potential energy depends not only on the coordinates q, but also on some parameters $r_1, r_2, \cdots r_m$, which may for instance be outside centers of force that exert external forces on the molecules of the gas.[178]

The same then holds also for the total energy

$$E = E(q, p; r_1, r_2, \cdots)$$

and consequently also for the right-hand side of the Hamiltonian equations (22) which determine the flow of G-points in the Γ-space.

Two different assumptions can be made with regard to parameters $r_1, r_2, \cdots$:

1. They are given arbitrary values, which, however, do not change with time. In this case the streamlines of the system in Γ-space remain unchanged with time, and the density distribution given in Eq. (53) will be stationary.

2. The parameters are arbitrary functions of time: $r_1(t), r_2(t), \cdots$. In this case the direction of flow in each point in Γ-space changes continuously and the density distribution of Eq. (53) will not be stationary.

22. Relations between average values for canonically distributed ensembles of systems

Let us first take assumption (1). Under this assumption Gibbs considers a few special phase functions

$f(q, p)$ and forms their average values over the canonical ensemble defined by Eq. (56). He determines mainly

A) The dispersion which such a function f has around the average value $\bar{f}$ within the canonical ensemble;

B) The dependence of the average values $\bar{f}$ on the modulus Θ and on the parameters $r_1, r_2 \cdots$.

The relationships thus established by Gibbs can be derived directly from the defining equations of these average values. They are based on the choice of the weighting function $\exp \{(\Psi - E)/\Theta\}$ that is used. We should note for the discussion of question (B) that the weighting function, i.e., the canonical density distribution, depends (1) on Θ explicitly; (2) through E on the $r_1, r_2, \cdots$; (3) through Ψ on Θ and on the $r_1, r_2, \cdots$. Let us present these results to the extent necessary for further discussion.

22a. Some of Gibbs's results. 1. Taking into account the fact that the number of degrees of freedom in a gas model is of the order of 10^{18} Gibbs proves the following theorem:[179]

(XII) In an ensemble which is canonically distributed with the modulus $\Theta = \Theta_0$, an overwhelming majority of the individuals will have very nearly the same total energy $E = E_0$.

2. The kinetic energy of each molecule (e.g., that of the k-th) is a positive definite quadratic form of its momenta $p_1^k, \cdots, p_r^k$. It can always be represented as

$$\frac{1}{2} \sum_{i=1}^{r} (u_i^k)^2,$$

where the $u_1^k \cdots u_r^k$ are r linear, homogeneous, and real combinations of the momenta. These combinations $u_1^k \cdots u_r^k$ (the so-called "momentoids"[180]) can be formed in infinitely many ways. Gibbs proves the following equality about the mean squares of $u_1^1 \cdots u_r^N$ formed over the canonical ensemble

$$(57)\qquad \overline{\frac{1}{2}(u_1^1)^2} = \cdots = \overline{\frac{1}{2}(u_s^k)^2} = \cdots = \overline{\frac{1}{2}(u_r^N)^2} = \frac{\Theta}{2}$$

from which one obtains the following simple relationship between the average kinetic energy of the whole gas model over the ensemble and the modulus θ:

$$(58)\qquad \overline{L} = \frac{rN\Theta}{2}\,.$$

3. The gas model exerts a certain reaction force at all times on the above-mentioned centers of force or on the piston. In a certain phase (q, p) of the motion it acts "along the parameter r_h" with the generalized force

$$(59)\qquad R_h(q, p; r) = -\frac{\partial\Phi}{\partial r_h} = -\frac{\partial E}{\partial r_h}$$

in the outward direction. Differentiating Eq. (55) with respect to r_h and combining it with Eq. (59), we get the following expression for the average of R_h over the ensemble:

$$(60)\qquad \overline{R_h} = -\frac{\partial\Psi}{\partial r_h}\,.$$

22b. Relationship to the Maxwell-Boltzmann distribution. In order to survey this relation and analogous connections as Gibbs develops them, it is necessary to touch upon a question which actually goes beyond Gibbs's treatment.[181]

Let $f(q, p)$ be a phase function that does not change its value upon the interchange of molecules. It will therefore assume a definite value if the Maxwell-Boltzmann distribution (Eq. 46) with $E_0, r_1, \cdots, r_m$ holds for the molecules of the gas. This value will be denoted by

$$(61)\qquad [f(q, p)]_{E_0, r}^{\mathrm{M.B.}} = B(E_0, r_1, \cdots, r_m).$$

Let us form furthermore

$$\overline{f(q, p)} = g(\Theta_0, r_1, \cdots, r_m), \tag{62}$$

the average value, over that canonical ensemble of gas models whose modulus precisely fits [cf. (XII)] the energy E_0. Now we may ask: What is the relationship in this case between quantities (62) and (61)?

According to (XII) it is first of all plausible that in general the average (Eq. 62) over the canonical ensemble will be very nearly identical with the average value taken over a microcanonical or even ergodic ensemble with $E=E_0$. In fact, in that case also Eq. (57), for example, coincides with a relationship derived by Boltzmann (1871) for ergodic ensembles.[182] Furthermore, the microcanonical ensemble is very nearly equivalent to an ensemble that is distributed (cf. Section 12c) with constant density over the "shell" in Γ-space belonging to

$$E(\mathbf{Z}) \equiv \sum a_i \epsilon_i = E_0.$$

If we now supplement Gibbs's discussion with the investigations of Boltzmann as presented in Section 13(I), we come to the following conclusion: In a canonically distributed ensemble of gas models the overwhelming majority of the individual members are in a state described by the Maxwell-Boltzmann distribution given in Eq. (46) with the parameters $r_1, \cdots, r_m$, and with the energy $E=\overline{E}$.

Thus, from the point of view of Boltzmann's presentation, the introduction of the canonical distribution seems to be an analytical trick reminiscent of Dirichlet's "discontinuous factor."[183] In the calculation of the average value $\overline{f(q, p)}$ the integrations (see Eq. 56) always remain extended over the infinite Γ-space. However, if the modulus Θ and the parameters $r_1, \cdots, r_m$ are chosen in the proper way, the decisive majority of all gas models will lie in those parts of Γ-space given by that M.-B.

distribution which belongs to the given E_0 and to the other prescribed conditions.

22c. Gibbs's measure σ for the deviation from the canonical distribution. As an aid in the investigations of nonstationary ρ distributions, Gibbs introduces a function as a measure for the deviation of a given ρ distribution from a canonical distribution with the same $\bar{E}$. This function, denoted by σ, is defined by[184]

$$(63) \qquad \sigma = \int_{-\infty}^{+\infty} \cdots \int_{-\infty}^{+\infty} \rho \log \rho dq_1^1 \cdots dp_r^N .$$

Its character as a measure is indicated by theorems of the following sort:

(XIII) Let us distribute a given set of G-points over a given region of the Γ-space. Then σ assumes its smallest value if the G-points are distributed with a constant density over the whole region in question.

(XIIIa) If the total set of G-points lying in each energy region E, $E+\delta E$ is given, σ assumes its smallest value if ρ is constant along each of the various energy surfaces.

(XIV) If we consider an ensemble with a total measure equal to 1 for which the average value $\bar{E}$ is given, the canonical ensemble belonging to this $\bar{E}$ will give the relatively smallest value of σ.

In proving statements (XIII) and (XIIIa), we use in the definition of σ only the "concavity" of the integrand $\rho \log \rho$. Thus the same conclusions also hold, e.g., for the integrand ρ^{2n}. However, in order to distinguish the canonical ensemble from other ensembles with the supplementary condition mentioned in (XIV), it was necessary to construct σ in a special way (Eq. 63).

The preparatory results, like Eqs. (57), (58), and (60) or statements (XII)–(XIV) which we have presented so far, were derived by Gibbs with elementary and completely explicit transformations. In fact, all that has been done so far is to derive the immediate consequences of the special choice (Eq. 53) of ρ.

However, as soon as we come to the main object of Gibbs's book, i.e., to the nonstationary ensembles, we meet with rather sketchy arguments. The reason for this is that only here do we have to begin to take into account the dynamical character of the statistical ensemble in question.

23. Nonstationary distributions of density in Γ-space

23a. The "disarrangement" of nonstationary distributions. First of all we have to analyze the assertions developed in Chapter XII of the book. They refer to the following question: What will become, as time goes on, of a $\rho(q, p)$ distribution which does not have the form (52) and hence cannot remain stationary?

Let us divide the Γ-space somehow into very small, but finite cells Ω: $\Omega_1, \Omega_2, \cdots, \Omega_\lambda, \cdots$, which might be, for instance, cubes of equal size. The average value which the "fine-grained" density $\rho(q, p, t)$ has at time t over the cell Ω_λ we will call "coarse-grained" density $P_\lambda(t)$ (read: capital ρ) of this cell. Because of Eq. (54) we have

$$\sum P_\lambda(t) = 1. \tag{64}$$

One can now formulate the statement at which Gibbs arrives at this point and which is fundamental for the whole subsequent development:

(XV) Every nonstationary fine-grained density distribution will be "disarranged" by the (stationary) streaming in Γ-space in such a way that the corresponding coarse-grained densities gradually assume stationary values.

H. A. Lorentz[185] characterized this tendency toward

a stationary P-distribution even better by the following supplemental statement:

(XV′) The limiting values of P are constant along the individual surfaces of constant energy; i.e.,

$$\lim_{t=+\infty} P(t, q, p) = F(E). \tag{65}$$

In order to support this statement, Gibbs uses the picture of the mixing of a nondiffusive dye in a colorless solvent. He gives in addition a more mathematical proof, which is reproduced by Lorentz in a more transparent way.

Remarks on (XV) and (XV′): 1. With a view to the analogies with thermodynamics to be discussed later, it is important to know the order of magnitude of the time needed for the approach to these stationary P-values. Let us take some reasonable value for the total energy of the gas model and let us consider first the time which, on the average, is needed for a G-point starting from a cell Ω_λ to return to this cell. Obviously we deal here with time intervals similar to those "enormously large" Poincaré-Zermelo cycles which we have already encountered in treating the *Wiederkehreinwand* (Section 7b). Hence if, in our support for statements (XV) and (XV′), we (like Gibbs and Lorentz) do not consider any other circumstances except those which are relevant in the mixing of a dye, then we can expect the approximate stationary state of P and the approximate realization of the P-distribution (Eq. 65) only after many Poincaré-Zermelo cycles. 2. Statement (XV′) and some of the other assertions to be discussed later obviously have a close connection with the ergodic hypothesis.[186] 3. The mathematical analysis given by Gibbs and Lorentz for the mixing process is incomplete in one essential respect: it uses the term "density" alternately to mean ρ or P. Thus we can understand how such a calculation can give so simply and without any special assumptions a mixing which increases with time.[187]

From Liouville's theorem, Eqs. (26) and (26′), it follows immediately that the quantity σ, which determines the distribution of the fine-grained density ρ, remains exactly constant during the mixing process. However, the function

$$\sum = \sum P_\lambda \log P_\lambda, \tag{66}$$

derived from the coarse-grained density P in an analogous way, can very well change with time.[188] Gibbs attempts, simultaneously with statement (XV), to prove the following statement:[189]

(XVI) Every nonstationary ρ-distribution will be disarranged by the stationary flow existing in the Γ-space in such a way that the following inequality will hold for the measure $\sum$ of inhomogeneity of the P-distribution:

$$\lim_{t=+\infty} \sum (t) \leq \sum (t_0). \tag{67}$$

Remarks on (XVI): 1. Gibbs explicitly emphasizes that he tries to prove only something about the lim $\sum(t)$ for $t=+\infty$, but does not want to assert that $\sum(t_2) < \sum(t_1)$ if $t_2 > t_1$.[190] 2. It is decisive for the later applications how low this limit turns out to be. If we admit statement (XV′) and combine it with theorem (XIIIa) applied to $\sum$ instead of σ, we obtain for the limit a value which still arbitrarily exceeds the value belonging to the corresponding canonical ρ-distribution. 3. Remark (1), if applied to (XV), is identical with the question, How much time will elapse before $\sum(t)$ noticeably attains its limiting value?

23b. The behavior of special nonstationary ensembles of gas models. In order to exhibit analogies with theorems of thermodynamics Gibbs proceeds to treat the behavior of certain special nonstationary ensembles.

He deals mainly with the following two problems:

A) Let us assume that up to time t_A all parameters $r_1, r_2, \cdots$, have the constant values $r_1^A, r_2^A, \cdots$, and and that up to this time the canonical distribution belonging to these parameters and to module Θ^A is

$$\rho^A = e^{(\Psi^A - E^A)/\Theta^A}. \tag{68}$$

From t_A on let us impose an arbitrary time variation on these parameters $r_1, r_2, \cdots$. The problem is to describe the subsequent distributions of phase.

B) Let us consider the case in which the molecular system is divided into two parts, part I and part II. These two parts can interact but only in such a way that the total energy E can be calculated additively from the energies of the two parts, i.e.,

$$E = E_{\mathrm{I}} + E_{\mathrm{II}}. \tag{69}$$

Let us prescribe the following distribution of phase for t_A:

$$\rho^A = e^{[(\Psi_{\mathrm{I}} - E_{\mathrm{I}})/\Theta_{\mathrm{I}} + (\Psi_{\mathrm{II}} - E_{\mathrm{II}})\Theta_{\mathrm{II}}]} \tag{70}$$

and let the parameters $r_1, r_2, \cdots$, be kept permanently constant. (One easily shows that this distribution, in general, cannot remain stationary.) The problem is to describe the subsequent distributions of phase.

Gibbs's relevant results could perhaps be formulated in the following way:[191]

(XVII) If the change of parameters mentioned in (A) consists of a discontinuous transition to the values $r_1^B r_2^B, \cdots$, which from then on stay constant, then the P-distribution, which will be reached after a very long time, is such that for the purpose of forming average values it can very nearly be replaced by the canonical ρ-distribution,

$$\rho^B = e^{(\Psi^B - E^B)/\Theta^B}. \tag{71}$$

(XVIII) If the change of the parameters $r_1, r_2, \cdots$,

in (A) occurs infinitely slowly, the P-distribution changes in such a way that it can always be approximately replaced by canonical ρ-distributions. During this process $\sum$ remains very nearly constant.

(XIX) In the situation described in (B), the P-distribution which is reached for $t = +\infty$ can also be approximately replaced by a definite canonical ρ-distribution.

Remarks on (XVII)–(XIX): Gibbs tries to prove these statements by combining (XV) and (XVI) with the minimum theorem (XVI) transcribed from σ to $\sum$. Keeping in mind, however, remark (2) to (XVI), we realize the following: While in all these cases Chapters XI and XIII prove at most a certain change in the direction of the canonical distribution, in Chapter XIV the analogies with thermodynamics are discussed as if it had been proved that the canonical distribution will be reached, at least approximately, in time. Such a jump can very easily remain hidden between the elementary but lengthy transformations from one inequality to the other.[192] But only with this jump is it possible to get from the generally qualitative considerations of Gibbs about nonstationary ensembles to the quantitative statement that in very many cases the final distribution is almost canonical.

We will accept statements (XV)–(XIX) in our subsequent discussions.

24. The analogy to the observable behavior of thermodynamic systems

We will compare the treatments by Gibbs and by Boltzmann of a few special but typical problems. In order to show clearly the conceptual contrast between the two treatments, it is helpful to emphasize the similarity of their mathematical apparatus. We will first list a few auxiliary formulas.

24a. Auxiliary formulas. The Maxwell-Boltzmann distribution of the phase points of the molecules in μ-space (Eq. 46) can obviously be written in the following form:

$$a_i = e^{(\psi - \epsilon_i)/\theta}, \tag{53'}$$

where the energy ϵ_i which the molecule has if its phase point lies in the ith μ-cell can depend on the parameters $r_1, r_2, \cdots$, we introduced above. As far as θ is concerned, there are two possibilities. Either it is actually given, in which case we can get ψ from the requirement that

$$\sum_i a_i = \sum_i e^{(\psi - \epsilon_i)/\theta} = N \tag{55'}$$

and the total energy E is then determined by

$$E = \sum_i a_i \epsilon_i = \sum_i \epsilon_i e^{(\psi - \epsilon_i)/\theta}. \tag{72}$$

Or, conversely, if the total energy is given, we can determine not only ψ but also θ from Eqs. (55′) and (72). The force which the N molecules in this distribution exert outward in the direction of parameter r_h is given by

$$R_h = \sum_i a_i \left(- \frac{\partial \epsilon_i}{\partial r_h} \right). \tag{59'}$$

By differentiating Eq. (55′) with respect to r_h, we can transform Eq. (59′) to:

$$R_h = - \frac{\partial}{\partial r_h} (N\psi) \tag{60'}$$

The calculation which Gibbs uses to derive Eq. (57) seems to be an adaptation of the calculation which Boltzmann (1871) used to get from Eq. (53′) the following relation:[193]

$$\frac{1}{N} \sum_{k=1}^{N} \frac{1}{2} (\overset{k}{u_1})^2 = \frac{1}{N} \sum_{k=1}^{N} \frac{1}{2} (\overset{k}{u_2})^2 = \cdots \frac{1}{N} \sum_{k=1}^{N} \frac{1}{2} (\overset{k}{u_r})^2 = \frac{\vartheta}{2}. \tag{57'}$$

From this we get for the total kinetic energy of the N molecules

$$L = \frac{N\vartheta r}{2}. \tag{58'}$$

If we now make the transition from the Maxwell-Boltzmann distribution (Eq. 53_1) belonging to $\vartheta, r_1, \cdots, r_m$ to that belonging to $\vartheta+\delta\vartheta, r_1+\delta r_1, \cdots, r_m+\delta r_m$, we get from Eq. (55_1) the following identity for these variations:

$$\sum_i e^{(\psi-\epsilon_i)/\vartheta}\delta\left(\frac{\psi}{\vartheta}\right) + \sum e^{(\psi-\epsilon_i)/\vartheta}\epsilon_i\frac{\delta\vartheta}{\vartheta^2}$$
$$- \sum_i e^{(\psi-\epsilon_i)/\vartheta}\frac{1}{\vartheta}\left(\frac{\partial\epsilon_i}{\partial r_1}\delta r_1 + \cdots + \frac{\delta\epsilon_i}{\delta r_m}\delta r_m\right) = 0,$$

from which we get with the help of Eqs. (55′), (72), and (59′) the following relation:

$$N\delta\left(\frac{\psi}{\theta}\right) + \frac{E}{\theta^2}\delta\theta + \frac{1}{\theta}(R_1\delta r + \cdots + R_m\delta r_m) = 0. \tag{73'}$$

If we make the transition from the canonical distribution (Eq. 53) belonging to $\Theta, r_1, \cdots, r_m$ to that belonging to $\Theta+\delta\Theta, r_1+\delta r_1, \cdots, r_m+\delta r_m$, we similarly get from Eqs. (55) and (59) for these variations

$$\delta\left(\frac{\psi}{\Theta}\right) + \frac{E}{\Theta^2}\delta\Theta + \frac{1}{\Theta}(\overline{R}_1\delta r_1 + \cdots + \overline{R}_m\delta r_m) = 0. \tag{73}$$

Having completed these formal preparations, we can now describe the typical analogies to the observable behavior of thermodynamic systems.

24b. Gas in thermal equilibrium and the equalization of temperature of two bodies of unequal temperature. 1. Interpretation of the behavior of the gas in thermal equilibrium at a given total energy E_0 and at given $r_1, \cdots, r_m$.

Boltzmann: One considers in Γ-space the "shell" of Γ-points which correspond to the given total energy E_0. The overwhelming majority of these phase points correspond closely to a Maxwell-Boltzmann distribution (Eq. 53′) of the molecules of the gas model (cf. Section 13, I). Then from Eqs. (57′), (60), etc., one calculates for this distribution of state the pressure and the other reactive forces, the kinetic energy per molecular degree of freedom, etc.

Gibbs: Using Eqs. (57), (60), etc., one calculates the average values, $\overline{R}_h$, $\overline{L}$, etc., for the canonical ensemble of gas models[194] belonging to $\overline{E}=E_0$ and to the given $r_1, \cdots, r_m$.

2. Interpretation of the equalization of temperature by the contact of two bodies K_{I} and K_{II} of unequal temperature.

Boltzmann:[195] Before the interaction the molecules of K_{I} and K_{II} will have the Maxwell-Boltzmann distribution corresponding to their initial temperatures. Hence, if we consider the interaction systems as one molecular system, the initial state is given. The average and also the overwhelmingly most frequent course of the subsequent group of motions will be the one obeying the *Stosszahlansatz* and the *H*-theorem (see Section 14c). As a consequence K_{I} and K_{II} will eventually assume Maxwell-Boltzmann distributions with identical ϑ. By Eq. (57′) this means that the kinetic energy per molecular degree of freedom is the same for K_{I} and K_{II}.

Gibbs: Before the interaction K_{I} and K_{II} will each be represented by a canonically distributed ensemble of gas models with moduli Θ_{I}^{A} and Θ_{II}^{A} respectively. In order to represent their contact we consider the ensemble to consist of every individual of ensemble I (in its instantaneous state of motion) interacting with every individual of ensemble II. We obtain in this way a $(2r_{\text{I}}N_{\text{I}}+2r_{\text{II}}N_{\text{II}})$-dimensional Γ-space and in it a density distribution which is given at the initial time t^A by formula (Eq. 70). If at the beginning we already had $\Theta_I^A = \Theta_{II}^A$, then the combined ensemble is canonical to start with and will remain stationary.[196] If, however, $\Theta_I^A \neq \Theta_{II}^A$, then in general the initial distribution will not remain stationary. It should therefore become disarranged according to statements (XV), (XVI), and (XIX), and the final canonical ρ-distribution can again be regarded as a joint distribution of two canonical ensembles I and II which by now have the same modulus, $\Theta_I^\infty = \Theta_{II}^\infty = \Theta^\infty$.[197]

Because of Eqs. (57) and (58) the average values over the ensemble of the kinetic energy per molecular degree of freedom is the same for K_I and K_{II}.

24c. The temperature as an integrating factor: The meaning of entropy and the increase of entropy for irreversible processes. 1. The temperature as an integrating factor for the quantity of heat in reversible processes: the meaning of entropy.

Boltzmann: If the external interaction with the gas model is infinitely slow, we may calculate as if the molecules had at each time a Maxwell-Boltzmann distribution corresponding to the instantaneous values of $E_1, r_1, \cdots, r_m$. With this assumption we get for the sum of the energy increase and the work performed for an infinitesimal transition, i.e., for the "supplied heat,"

$$\delta Q = \delta E + (R_1 \delta r_1 + \cdots + R_m \delta r_m), \tag{74'}$$

which, using relationship (73′) can be transformed into

$$\frac{\delta Q}{\theta} = \delta\left(\frac{E - N\psi}{\theta}\right). \tag{75'}$$

The structure of the quantity on the right-hand side of Eq. (75′), which plays the role of the entropy, immediately leads to the following representation:

$$(76') \qquad \frac{E - N\psi}{\theta} = - \sum_i \log \left(e^{(\psi-\epsilon_i)/\theta}\right) e^{(\psi-\epsilon_i)/\theta}.$$

This representation in turn serves as an incentive to introduce tentatively, as a generalization of entropy, the function

$$(77') \qquad -H = - \sum_i a_i \log a_i$$

for any arbitrary distributions of state a_i.

Gibbs: If one affects all the members of an initially canonical ensemble in a manner corresponding to the reversible change of the state of a gas,[198] then it is permissible to assume (cf. XVIII) that the ensemble will always pass through canonical distributions only. Under these assumptions we can represent the average value over the ensemble of the δQ defined above by[199]

$$(74) \qquad \overline{\delta Q} = \delta \overline{E} + (\overline{R}_1\, \delta r_1 + \overline{R}_m \delta r_m),$$

which then can be transformed with the help of Eq. (73) into

$$(75) \qquad \frac{\overline{\delta Q}}{\Theta} = \delta\left(\frac{\overline{E} - \Psi}{\Theta}\right).$$

The following representation can then be made:

$$(76) \qquad \frac{\overline{E} - \Psi}{\Theta} = -\int_{-\infty}^{+\infty} \cdots \int \log \left(e^{(\psi-E)/\Theta}\right) e^{(\psi-E)/\Theta} d\overset{1}{q_1} \cdots d\overset{N}{p_r},$$

and correspondingly Gibbs's attempts to define the quantity

$$(77) \qquad -\sigma = -\int_{-\infty}^{+\infty} \cdots \int \rho \log \rho\, d\overset{1}{q_1} \cdots d\overset{N}{p_r}$$

or the quantity $(-\sum)$ as the entropy even for arbitrary ρ-distributions. This shows in detail the formal analogy with Boltzmann's treatment of the H-function.

2. The increase of entropy for irreversible processes in as isolated system.

Boltzmann: If the phase points of the molecules in μ-space at time t^A is not of the Maxwell-Boltzmann type, then the quantity H which controls the a_i-distribution will for the overwhelming majority of the motions after t^A assume smaller values than it had at t^A.	Gibbs: If the phase points of the gas in Γ-space at time t^A is neither canonical nor of the general type (Eq. 52),[200] then the quantity $\sum$ which controls its P_λ-distribution will, after an infinitely long time, assume a smaller value than it had at t^A.

24d. Remarks on the interpretation of entropy by means of Gibbs's measure $(-\sum)$. 1. M. Planck emphasizes the following point:[201] In constructing the quantity H, Boltzmann from the beginning considered the question of whether all molecules are the same, i.e. (see Sections, 12b and d), whether they can be permuted among themselves, or whether one has a mixture of gases. This circumstance, however, is not considered at all in the definition of the quantity $\sum$. Consequently Gibbs's definition gives "no information about the way the concentrations of the various types of molecules influence the additive constant in the expression for entropy." Planck indicates how one should fill in this gap by the introduction of another function. L. S. Ornstein,[202] on the other hand, finds that the $(-\sum)$ for a gas model containing several types of molecules is in complete agreement with the thermodynamically defined entropy of a mixture of gases in equilibrium. It should be noted, however, that he, deviating from Gibbs's method, has to resort to the permutability of molecules of the same type (as in Boltzmann's treatment) when he calculates the increase of entropy in the diffusion of two gases.

2. H. A. Lorentz[203] explains why the quantity $(-\sum)$ cannot be considered a satisfactory interpretation of the entropy when we deal with nonstationary cases, and he points out the way to establish a more satisfactory interpretation.

3. Relationship to the *Wiederkehreinwand.* One cannot prove a quasiperiodic behavior of quantity $\sum$ similar to that of quantity H. This is so because, while Poincaré's theorem is based above all on the assumption of a finite number of degrees of freedom, $\sum$ is a measure of the disarrangement of a continuum of G-points in Γ-space. It might appear, therefore, that Gibbs has eliminated the *Wiederkehreinwand* by interpreting the entropy of the gas by $(-\sum)$ instead of $(-H)$. However, consideration of one function or the other can obviously not alter the fact that the single G-point of the ensemble describes a cycle on its E-surface in Γ-space, i.e., that the single gas model performs its Poincaré-Zermelo cycle. We can say even more. According to the developments of Chapter XII (as we have already pointed out; see remark 1 on XV and XVI′), the disarrangement of the nonstationary initial distribution of P, and with it the decrease of $\sum$ to a relative minimum, are in essence caused precisely by the repeated cyclic motions of the single G-points.

A more detailed comparison shows that the superiority of Boltzmann's theory regarding the increase of entropy is also based in this case on the proper consideration of the permutability of the molecules.

If one would work out the suggestions of Planck and Lorentz, then Gibbs's discussion of the decrease of $\sum$ will probably become essentially an investigation of the decrease of the average values of H over the ensemble (cf. Section 14c).

24e. The monocycle analogies with thermodynamics. (Cf. V 3.) When these analogies were pointed out by Helmholtz (1884), J. J. Thomson, and others, Boltzmann investigated their origin from the point of view of the kineto-statistical theory. These papers by Boltzmann (1884, *Collected Works*, III, Nos. 73, 74, 82) should be mentioned at this point because he treated in them exactly the same methods and thermodynamical analogies that were emphasized fifteen years later by Gibbs in his treatment of the microcanonical ensemble. In the papers quoted Boltzmann starts from the consideration of an ergodically (i.e., microcanonically) distributed ensemble. He changes this distribution infinitely slowly by varying infinitely slowly certain parameters $r_1, \cdots, r_n$ contained in the potential energy. In this way Boltzmann arrived at a profound criticism of the monocycle analogies which has not been considered in any of the accounts of Helmholtz' monocycle theory.

25. Articles following or related to Gibbs's treatment

The articles of L. S. Ornstein (1908–1909) are, first, very valuable contributions to the clarification of the foundations of Gibbs's treatment. In addition, they show (1) that operations with canonical ensembles occasionally furnish a more convenient computational scheme for the treatment of complicated problems of equilibrium (e.g., the equilibrium in capillary transition layers) than does Boltzmann's procedure; (2) the reason why the two methods give the same result in a large group of equilibrium problems.

A. Wassmuth (1908) shows that among all distributions of the form $\rho = F(E)$ only the canonical distribution satisfies the following requirement: Let us consider only those G-points of the ensemble which give a certain definite configuration $(q_1^1, \cdots, q_r^N)$ to the molecules of the gas model for arbitrary values of the velocities. Now let us form for these particles the average of the square of a momentoid (see note 179). We require that this average be independent of the prescribed configuration.

In order to characterize the contribution of some of the other papers, it is useful to recall a consequence which Boltzmann (1871) derived from the ergodic hypothesis and from Eq. (34).[204] It is actually the origin of the idea of representing the behavior of a body in thermal equilibrium by the average behavior of a canonical ensemble. In an ergodic system consisting of N molecules, let us consider a group of N' molecules, where N' may be a large number but still very small compared to N. Boltzmann obtains an expression for the relative length of time during which the state of these N' molecules lies in the region $dq_1^1 \cdots dp_r^N$. This expression is

$$dW = ce^{-E'/\Theta} dq_1^1 \cdots dp_r^{N'} \tag{78}$$

where E' is the total energy of the group of molecules in this state and $\Theta/2$ the time average of the kinetic energy per degree of freedom of the ergodic system. If we consider instead the corresponding stationarily distributed ergodic ensemble (Eq. 31a), then Eq. (78) will be proportional to the number of such individuals in the group for which the state of the N' molecules under consideration lies in the region $dq_1^1 \cdots dp_{1r}^{N'}$. This is the form in which one finds the theorem in Maxwell's work (1878) [3]. Gibbs expresses it in the following way (*op. cit.*, p. 183): "If a system of a great number of degrees of freedom is microcanonically distributed in phase, any very small part of it may be regarded as canonically distributed." The part of the system of N' molecules in whose behavior we are interested is the body, while the whole ergodic system is this body together with a very large temperature bath. This is the way in which Einstein also uses the ergodic hypothesis and the microcanonical and canonical ensembles in two papers on the "kinetic theory of thermal equilibrium and the second law of thermodynamics" (1902, 1903) [1, 2]. He employs the ensemble function $(-\sum)$ for the interpretation of the entropy of a body and the second law is interpreted correspondingly.

The expression given in Eq. (78), however, can be used in other ways than to obtain information about the average behavior of the group of N' molecules. It tells us with what relative frequency the states different from the average behavior occur in a given group of molecules during a sufficiently long period of time, or—by use of the canonical ensemble corresponding to assertion (78)—with what relative frequency the various states are represented in it. Therefore, once we accept the ergodic hypothesis, Eq. (34), and the transition to assertion (78), we can use this canonical ensemble as a formal tool for solving all those problems dealing with the "probability of a certain deviation from the most probable state."

M. von Smoluchowski (1903)[205] studied this problem of "irregularity" from several angles in connection with the following problem: Consider the deviations from the "most probable" spacial distribution which the molecules of a gas in thermal equilibrium show at various times. What is the effect of these deviations on the equation of state and the H-value? Smoluchowski in this paper referred to the relationship of this problem to the stability limit for a superheated fluid and supercooled vapor, which he investigated more closely in a later paper (1907).[206]

The period between these two dates saw the publication of two papers by Einstein (1905)[207] and of two by Smoluchowski (1906)[208] on the Brownian motion. In the present context the reasoning of Einstein's second paper is of special interest. Let us assume that the N' molecules discussed in connection with the expression given Eq. (78) compose a microscopically small particle suspended in a liquid which is in thermal equilibrium. In order to determine the instantaneous state of motion of the particle, we will use data like the height of the center of gravity or the angles characterizing the orientation of the particle as well as the coordinates of the internal configuration and motion. One of these latter parameters we will denote by α. Furthermore, we will assume that the potential energy of the particle contains the parameter α only in an additive term $\chi(\alpha)$. Integrating Eq. (78) over all the parameters except α, we obtain for the relative time during which α lies between α and $\alpha + d\alpha$, an expression of the form

$$dw(\alpha) = c \cdot e^{-\chi(\alpha)/\theta} d\alpha, \tag{79}$$

where c is independent of α. Let us assume that $\chi(\alpha)$ has a minimum at $\alpha = \alpha_0$. If thermal motion were absent, the particle would be in equilibrium at $\alpha = \alpha_0$. Eq. (79) therefore gives information regarding the deviations of this parameter α from the most probable value α_0 due to thermal motion. We emphasize two of the applications of Eq. (79) made by Einstein:[209] (1) to the distribution of microscopic suspensions in a liquid under the action of gravity; (2) to the mean square of the changes of parameter α due to thermal motion in a certain time interval τ (specifically; the horizontal wandering and rotation of the suspended particles).

26. Conclusion

The foregoing account dealt chiefly with the conceptual foundations of statistico-mechanical investigations. Accordingly we had to emphasize that in these investigations a large number of loosely formulated and perhaps even inconsistent statements occupy a central position. In fact, we encounter here an incompleteness which from the logical point of view is serious and which appears in other branches of mechanics to a much smaller extent. This incompleteness, however, does not seem to have influenced the physicists in their evaluation of the statistico-mechanical investigations. In particular, the last few years have seen a sudden and wide dissemination of Boltzmann's ideas (the *H*-theorem, the Maxwell-Boltzmann distribution, the equipartition of energy, the relationship between entropy and probability, etc.). However, one cannot point at a corresponding progress in the conceptual clarification of Boltzmann's system to which one can ascribe this turn of affairs.

It is much more likely that the study of electrons and the investigation of colloidal solutions with the ultra-microscope have been responsible. In general, both of these have had the effect of reviving and deepening the concept that all bodies can be pictured as aggregates of a finite number of very small and identical elementary components, and that correspondingly every process in a physical or chemical problem which can be observed by normal methods is a complex of an enormously large number of individual processes. The opportunity arose to apply the methods of the kinetic theory of gases to completely different branches of physics. Above all, the theory was applied to the motion of electrons in metals (V 14, Section 40, by H. A. Lorentz), to the Brownian motion of microscopically small particles in suspensions (Section 25), and to the theory of black-body radiation (V 23, by W. Wien).[210] In all these cases, however, we have to do with a tentative transference of the most important results without a corresponding development of the foundations; e.g., the Maxwell-Boltzmann distribution law and the theorem of the equipartition of energy is transferred without hesitation to the thermal motion of electrons in metals, although the interaction with the ether, which in the Maxwell-Boltzmann theory is completely neglected, must certainly play an important role here.

A justification of these statistical methods in the new fields of application is the fact that here some experiments allow a much more detailed testing of the results of these methods than was possible in the previous applications.[211] For instance, O. W. Richardson (1908–1909),[212] using a very direct electrical method, succeeded in measuring the velocity distribution and the average kinetic energy of electrons ejected from a heated metal by thermal motion. Concerning the results obtained by the microscopical investigation of the Brownian motion, we refer to the comprehensive reviews of Smoluchowski and Perrin.[213] In all these cases the measurements gave results which can be considered a very direct and extensive confirmation of the kinetic assumptions. How the differences[214] still remaining can be interpreted will, of course, be decided only by future extensions of these experiments.

The situation is more complicated in the applications of the statistical method to radiation phenomena. Some of the older attempts have not as yet yielded clear results. Among these are the connection of the interference limit for large path differences with the average time between collisions undergone by the center of emission,[215] or the remark that, on account of the corresponding Doppler effect, the thermal motion of the sources of emission creates a lower limit for the width of fine spectral lines.[216]

The theorem of the equipartition of energy when extended to the thermal equilibrium between matter and ether was very well confirmed as far as the infrared part of black-body radiation was concerned. Its extension to the ultraviolet domain, however, leads to absurd results, so that, at least for the time being, one is unable to derive the Boltzmann-Stefan law and Wien's displacement law without reference to thermodynamical results. At the present time one cannot see how these difficulties can be solved.[217]

Analogous difficulties are encountered in the gas theory itself.[218] While the law of equipartition of energy can very well be applied to the translational and in some cases to the rotational motion of molecules, it apparently fails for the complicated internal motion of molecules. On the other hand, this theorem is an immediate consequence of an assumption that underlies all of Boltzmann's investigations concerning thermal equilibrium, namely, that when forming average values in Γ-space we have to use those ensembles that are distributed with a constant density along the surfaces of constant energy.[219]

At least in this respect, therefore, a further development of the foundations of statistical mechanics has unquestionably become necessary.

Appendix

SINCE the closing date of this article, which reached the publisher in the above form, ready for print, in January 1910, a series of important articles have been published which take a stand especially with regard to the problems discussed in Sections 23–26. We will report on these papers in the form of appendixes to the sections dealing with the respective problems.

27. Appendix to Section 23: Nonstationary distributions of density in Γ*-space*

Since the treatment of Gibbs was incorrect,[220] Jan Kroò[221] has tried to give a more detailed analysis of the way in which a nonstationary density distribution through the streaming in Γ-space rearranges itself and approaches a stationary distribution. First he considers a periodic system with one degree of freedom. The two-dimensional space can then be represented (at least piecemeal) in such a way that the orbital curves are concentric circles. If the period for the various orbits is different, these circles rotate with different angular velocities.[222] Under the assumption that the fine-grained density distribution $\rho(q, p)$ is sufficiently continuous for $t = t_0$ and that the condition of stationarity (i.e., the constancy of ρ along the individual orbits) is violated, Kroò shows that in the limit of $t = +\infty$ the coarse-grained density P will be constant along the orbits and hence also constant in time. At the same time we will have

$$\lim_{t=+\infty} \sum (t) \leq \sum (t_0).$$

Then Kroò turns to periodic systems with n degrees of freedom where the period changes with at least one of the constants of integration $c_1, \cdots, c_{2n-1}$ which determine the G-path of the system (cf. Section 9b), e.g., with c_1. Let us classify the orbits into ∞^{2n-2} groups, each with ∞^1 orbits. Then the same kind of rearrangement or mixing up will take place on the two-dimensional surface corresponding to each group as occurred in the two-dimensional Γ-space of the previous example. Thus the statements made for the system with one degree of freedom will also hold for this more general case when we study the behavior of the coarse-grained density and the quantity $\sum(t)$ at $t=+\infty$.

The analysis of Kroò shows in fact that for $t=+\infty$ P lies closer to the ergodic distribution given in Eq. (30) than at $t=t_0$ and that similarly $\sum(t)$ lies closer to the corresponding $\sum$ value at $t=+\infty$.[223] However, it would be a mistake to confound this result, which corresponds to statement (XV) in Section 23d, with the assertion that for $t=+\infty$ the ergodic distribution and the corresponding $\sum$-value are approximately attained. The latter corresponds to Gibbs's indispensable statement (XV′). It is precisely for the periodic systems treated by Kroò that it is particularly easy to see that the transition from (XV) to (XV′) necessarily invokes an assumption similar to the ergodic hypothesis.[224] (Cf. the remarks in Sections 23a and 23b).

28. *Appendixes to Sections 24 and 25: The analogy to the observable behavior of thermodynamic systems* and *Articles following or related to Gibbs's treatment*

The investigations of P. Hertz try to clarify the physical meaning of Gibbs's procedure when he, for the treatment of the thermodynamic behavior of a single system, introduces the microcanonical or canonical ensembles of systems. First Hertz rejects[225] the canonical ensemble as a purely formal construction. On the other hand, because of the ergodic hypothesis, he assigns real physical significance to the microcanonical ensemble, i.e., to the ergodic surface density distribution (Eq. 31) as introduced by Boltzmann and Maxwell. Correspondingly he obtains Boltzmann's theorem for the equality of the average kinetic energy for all degrees of freedom. In order to establish its connection with the thermodynamic concept of thermal equilibrium he investigates more in detail the assumptions which are needed to derive the following two statements:

A) "Theorem of union": If two systems have the same kinetic energy per degree of freedom before they are in contact with each other, this will remain the case after the contact.

B) "Theorem of separation": If we consider a system which consists of two bodies in contact and we separate this system into its two constitutents, both of these parts will have the same kinetic energy per degree of freedom as the previously united system had.

The proof of the second theorem requires very far-reaching assumptions. Using Boltzmann's terminology, we have to make essentially[226] the following assumptions: The time interval during which the energy in the coupled system is distributed over the two parts very nearly according to the theorem of equipartition of energy is overwhelmingly larger than the time intervals for which this is not the case. For ideal gases one can show the validity of this assumption (still, of course, using the ergodic hypothesis) by using the volume of the corresponding Γ-regions. This leads to calculations which are essentially similar to Boltzmann's calculations discussed in Section 12b. For other, e.g., solid, bodies, we lack of course the means for carrying out a similar investigation. Using this as a basis, Hertz further discusses the analogies to thermodynamics.

In a second paper[227] P. Hertz, with an appeal to the investigations of L. Ornstein [1], assigns physical meaning also to the canonically distributed ensembles, first on the basis of considerations which depend on the ergodic hypothesis and which are therefore conceptually identical with the considerations of Boltzmann (1871), mentioned in Section 25 in connection with Eq. (78),[228] and second on the basis of the arguments which Gibbs in Chapter XIV of *Statistical Mechanics* (pp. 185–188 in the German edition) used to support the preference of canonical ensembles to microcanonical ensembles.[229] Let us consider a system from an empirical point of view, e.g., a body at a certain temperature. The energy of this system is not given exactly but only approximately. For this reason it might be sensible to describe the behavior of this system using the average distribution of an ensemble of systems which can assume (with rapidly decreasing probability) also the neighboring energy values.

L. Ornstein[230] discusses related questions. One should emphasize (1) his critical attitude toward the ergodic hypothesis and, as a result, his introduction of the microcanonical ensemble not as the only possible one, but only as the simplest stationary ensemble;[231] (2) some remarks on the properties of ensembles of systems which are distributed around the surface of most probable energy following a different law, e.g., with density proportional to

$$e^{-(E-E_0)^2k} \text{ instead of } e^{-(E-E_0)k}.$$

A. Einstein[232] discusses the relationship of the foundations of his presentation to that of Gibbs's *Statistical Mechanics*. P. Debye[233] in his paper on the electron theory of metals uses the scheme of Gibbs. In determining the measure for entropy, he employs a combinatorial procedure which operates with a distribution of a large, finite number of gas image points over Γ-space in the same way that Boltzmann does with the distribution of molecule image points over the μ-space.[234] In this process, strangely enough, he uses in the case of a mixture of gases the permutability of molecules of the same kind.[235] This is done with a view toward Planck's criticism (Section 24d) of Gibbs's assertions about the entropy.

A comprehensive study of the logical relationship between Boltzmann's definition of entropy, on the one hand, and the various proposals Gibbs has made,[236] on the other, is not yet available. For a system with very many degrees of freedom the various entropy definitions of Gibbs become equivalent to each other.[237] Also in this case all of these agree with Boltzmann's measure of entropy, inasmuch as the entropy of a state will be identical with the logarithm of the "probability" of this state.[238] Gibbs, however, characterizes the "state" only by the total energy E and the values of the parameters $r_1, \cdots, r_m$. Correspondingly the relative probability of two "states" will be measured by the relative number of phase points that give the prescribed total energy E at the given values of the parameters $r_1, \cdots, r_m$. In this process one counts indiscriminately not only those phase points which lead to thermal equilibrium, but also those which represent arbitrary large deviations from equilibrium. Boltzmann, on the other hand, characterizes the "state" of the gas in considerably more detail by giving in addition to the values of $r_1, \cdots, r_m$ the full distribution of state Z (Section 12a) which the molecules have. Correspondingly Boltzmann measures the relative probability of two states by the relative number of those phase points which satisfy this stricter characterization of state. It is therefore immediately clear why for states of equilibrium (in the study of reversible processes) Boltzmann's measure of probability and hence also his measure of entropy can be replaced by the probability and entropy measure of Gibbs. This is based on the fact that—as Boltzmann has shown (Section 13)—the overwhelming majority of all Γ-points correspond to thermal equilibrium. This replacement may indeed mean in many cases a substantial simplification of the calculations. On the other hand, it is also clear that Gibbs's measure of entropy is unable to replace Boltzmann's measure of entropy in the treatment of irreversible phenomena in isolated systems, since it indiscriminately includes the initial nonequilibrium states with the final equilibrium. It is possible to visualize a corresponding extension of Gibbs's presentation, e.g., by the inclusion of parameters of the distribution of state in the exponent of the canonical density distribution. This could be done in the same way that Gibbs brings in other requirements also on the "state" by adding appropriate terms in the exponent of the density ρ.[239]

29. Appendix to Section 26: Concluding remark

The restriction to ergodic distributions—as has been mentioned before—leads to serious contradictions with experience, since it immediately gives the theorem about the equality of the average kinetic energy for all degrees of freedom regardless of how unlike these degrees of freedom may be. These contradictions were recently strongly underlined by the investigations of W. Nernst and his students concerning the specific heat of solids at low temperatures.[240] The theoretical analysis of the spectral distribution of black-body radiation has already shown that, when an ether oscillator of high frequency competes with an ether oscillator of lower frequency, the latter will get, on the average, much more energy than the former. It is only for very high temperatures or very low frequencies that the oscillators satisfy in a reasonable degree the theorem of the equipartition of energy.[241] On the other hand, Einstein has pointed out[242] that the specific heat of solids seems to deviate more and more from the law of Dulong-Petit the lower the temperature is, and that it has a tendency to go to zero at the absolute zero of temperature. A connection between the two fields was established by Einstein by the assumption that the energy is distributed among the atoms that take part in the thermal oscillations in the same proportion as among the systems of ether oscillators of various frequencies. It is further assumed that the atoms of a given solid vibrate around their equilibrium positions with a characteristic frequency, which is determined by the atomic mass and the modulus of elasticity of the solid.[243] The measurements of Nernst which went down as low as 23° K have verified for a large group of materials the essential features of the parallel behavior of an ether and an atomic oscillator which Einstein had postulated. Above all, the statement that the specific heat of solids converges to 0 very strongly as the temperature approaches 0 was convincingly demonstrated.

The statistico-mechanical theory of thermal equilibrium can explain this fact only by considering, instead of the ergodically distributed ensembles of systems, such distributions in Γ-space for which the nonequivalence of degrees of freedom of different nature can properly be taken into account. The only detailed attempt in this direction is Planck's theory of black-body radiation,[244] and this was also carried over unchanged by Einstein to the thermal vibrations of solids. We need, however, more experimental and theoretical investigations to determine which are the nonergodic ensembles leading to the energy distributions realized in nature,[245] and for which among these the analogies with the second law and especially the relationship between entropy and "probability"[246] are preserved. From this point of view the trick which Einstein uses systematically deserves special attention. He retains the relationship:

Entropy = Logarithm of the "probability"

and then, reversing Boltzmann's procedure, he calculates the relative "probability" of two states from the experimentally determined values of the entropy. In this way he calculates—always appropriately adapting the meaning of "probability"—the average values, in time or otherwise, of the parameters of the state of a system. In those applications where we know from experience that there is a violation of the theorem of the equipartition of kinetic energy, this procedure goes essentially beyond the range of validity of the methods of Boltzmann.[247]

The experimental arrangements which can measure and in fact directly count (!) the statistical appearance of physical effects of average values have recently been considerably improved.[248] The extension of the statistical treatment to a steadily increasing range of physical phenomena[249] gives the "statistical experiment" an increasing methodological significance in the whole of physical research. A remark of E. von Schweidler (1905)[250] has contributed greatly to directing the interest of physicists toward the experimental determination of the relative frequencies with which a parameter deviates in various degrees from the "most probable" value. Von Schweidler points out that one can conclude from the dispersion of a parameter whether the physical effect in question is a result of the combined action of a finite number of identical and independent individual effects and how large their number is. This procedure is related to the method which is also used in population and biological statistics where—following W. Lexis and K. Pearson—the size of the dispersion is a criterion for the independence of the individual cases.[251] Some of the concepts and formal tools which have been worked out in these fields are still awaiting their application in the field of physical statistics.[252] Conversely one can expect that physical statistics will be a pioneer for other branches of statistics, because it excels in the relative simplicity of its individual phenomena, in the well-defined character of its experimental conditions, and, above all, in the ease with which it allows a great number of mass observations.[253]

30. Appendix to Section 19: The problem of axiomatization in kineto-statistics

A satisfactory characterization of the similarities and differences between the "probability hypotheses," on the one hand, and the usual hypotheses of natural sciences, on the other, has —apparently—so far not been given. The usual hypotheses of the natural sciences consist of the statement that a certain abstract system of defining axioms and theorems gives a picture (sufficiently accurate) of an actual phenomenon. The "probability hypotheses" give something similar; they also juxtapose a certain abstract scheme and an actual phenomenon. (The abstract scheme can be, e.g., a combinatorical dice scheme or an urn scheme.) The character of the representation in this case is, however, quite different. One places one actual phenomenon E_{I} against a whole ensemble S_{I} of different occurrences in the abstract scheme. How, therefore, can a "probability hypothesis" be verified experimentally? One can test if the actually observed outcome of E_{I} coincides sufficiently accurately with the "most probable" occurrence in the ensemble S_{I}. Furthermore one can consider the event E_{II} which consists of n_{I} repetitions of E_{I}, and one can see whether the outcome of E_{II} coincides sufficiently accurately with the "most probable" type of ensemble in the corresponding combinatorical ensemble S_{II} of ensembles S_{I}. Similarly one can consider the event E_{III} which consists of the n_{II}-fold repetitions of E_{II}, and the corresponding combinatorical ensemble S_{III} of ensembles S_{II}, etc. In other words, one can try to describe a certain physical phenomenon (e.g., the emission of α particles by radium) by "experiments and hypotheses of the first, second, third, . . . , kth order."[254] What is the logical relationship between hypotheses of the kth and of the $(k+1)$th order (H_K and H_{K+1})? In any case, the following fact is noteworthy: Let us assume that we observe the outcome of the experiment E_K, which from the viewpoint of hypothesis H_K is particularly unsatisfactory. This outcome, from the point of view of hypothesis H_{K+1}, is not only admissible, but hypothesis H_{K+1} requires that if we repeat the experiment $E_K n_K$ times, i.e., perform experiment E_{K+1}, this particular outcome should indeed occur with a very definite frequency. If one continues this kind of argument, one encounters difficulties[255] that can be masked in many ways but of which a satisfactory analysis has not been given. As long as we deal with the quite probable small deviations from the "most probable," the contrast between the "probability hypotheses" and the usual hypotheses of the natural sciences will hardly be discernible: both kinds of hypotheses will have the character of an approximation. The situation is different if the question of arbitrarily large deviations is seriously considered. Then the contrast becomes evident. The physicist will ask himself if he should allow for such large deviations from the "most probable" in the picture that he creates of this phenomenon. Boltzmann in one case committed himself to this[256] in a completely unreserved way: he explicitly admits cases where the entropy spontaneously decreases. Planck, on the other hand, decides in the opposite way in connection with the same case and, generalizing, emphasizes the following:[257] The physicist is free to exclude by a special physical hypothesis those deviations which would result in the violation of the uniqueness which is admitted for the macroscopic changes (in time) of a physical phenomenon. (Cf. Section 15.)[258] Usually the physicist will avoid making a definite decision. He is inclined simply to disregard the strong deviations from the most probable on the basis of their being so "enormously improbable." Or, even more generally, he will refuse to discuss such distant consequences of a physical theory. Until recently these questions emerged only on the horizon of physical research: in the theory of observational error[259] and in the discussion of the H-theorem. Now the situation has changed. The increasing infiltration of statistical elements in almost all physical concepts and the rising importance of the "statistical experiment" as a tool of physical research make these questions as urgent for the physicist as they already had been for some time for the theorist in population statistics, biological statistics, etc.[260] At present all investigations about the structure of a physical theory inevitably lead to the question of the nature of the "probability hypotheses."

Notes

PREFACE TO THE TRANSLATION

1. I expressed this point of view for the first time in a paper entitled "Zur Axiomatisierung des zweiten Satzes der Thermodynamik" (*Zeits. f. Phys.*, 33 and 34, 1925). A more complete discussion is contained in my book *Die Grundlagen der Thermodynamik* (Leiden, 1956).

2. Only a few authors take my views into account. The ones I know of are: J. D. van der Waals, *Lehrbuch der Thermostatistik*, 3d ed., ed. by P. Kohnstamm (Leipzig, 1927); A. Lande, *Axiomatische Begrundung der Thermodynamik* (Handbuch der Physik, vol. IX); and lately A. H. Wilson, *Thermodynamics and Statistical Mechanics* (Cambridge, 1957).

INTRODUCTION, TEXT, AND APPENDIX

1. The number of molecules in 1 cm^3 at 0° and atmospheric pressure (i.e., the Loschmidt number) is about 4×10^{19}. (Its value as known today is $2.68709 \pm 0.00009 \times 10^{19}$.—THE TRANSLATOR)

2. Cf. V 8, Sections 1 and 26 (L. Boltzmann and J. Nabl).

3. These "probability assumptions" are therefore statements about the statistical regularities among the billions of molecules which compose one gas quantum. The position and motion of one molecule is considered here the single event; the behavior of the one given gas quantum is a mass phenomenon. Hence occasionally these older works were also referred to as "statistico-mechanical" studies. Complying, however, with present-day terminology, we will reserve the term "statistico-mechanical" in our discussions to a group of investigations in which the behavior of one gas quantum is considered an individual event and where the behavior of a whole set, to be specified later, of infinitely many copies of such a gas

quantum is taken to be a mass phenomenon. All these copies are identical and move independently of each other. (Cf. Sections 9–15.)

4. V 8, Sections 15–25.

5. Boltzmann [6].

6. See Section 6 and V 8, Section 11.

7. Loschmidt [1].

8. Zermelo [1].

9. Section 7.

10. Sections 9–15.

11. Zermelo [3] (1900); Poincaré [3] (1908); Brillouin [1] (1902–1905); Lippmann [1] (1900); Liénard [1] (1903); Burbury, *Phil. Mag.*, (6) 16 (1908), 122.

12. Zermelo [1].

13. Cf. note 15.

14. The arrangement of the next eight sections parallels the treatment in V 8.

15. In the following discussion we will discard the term "the probability of an occurrence" and will replace it by the clumsier but explicit statements about frequencies. It is true that in the older work, which we will discuss now, the uncritical use of the term "probability" proved to be heuristically fruitful. Consider, for instance, the following two different physical quantities: (a) the relative time which a molecule A spends in state S, (b) the relative number of those molecules which at a given time are in state S. These two different frequencies were both described by the expression: "The" probability of A being in state S, and no further differentiation was made. In this way the identity of the two frequencies was implicitly assumed, and this identity was then applied in further considerations. The actual proof of the identity would perhaps have encountered great difficulties, if it could have been done at all. It is unfortunate that this procedure has not been completely eliminated even in the more recent critical studies of the foundations of the kinetic theory. Cf. the beginning of Section 14.

16. Cf. V 8, Section 2.

17. Krönig [1].

18. Because of the special form of the sums which appear in the calculation of the pressure, the Krönig model leads in this problem to the same results as the later, more-refined scheme of Clausius. In other problems, however, such as diffusion and heat conduction, this is no longer the case.

19. The following remarks refer primarily to those papers of Clausius [1, 2], which preceded the first publication of Maxwell [1], on the theory of gases.

20. This statement obviously formulates one of the many meanings in which the expression "the equal probability of all directions of the velocity" has been used. The same can be said about the phrase "the probability that an individual molecule is in a certain specified volume element of the container." Cf. note 15.

21. Cf. the explanation of this assumption applied to a simplified model in the Appendix of Section 5.

22. Our considerations will be limited, for the sake of simplicity, to spherical molecules. Then the "action sphere" of a molecule is a sphere concentric with the surface of the molecule but having a radius double that of the molecule. It marks the limit of the distance which the centers of two molecules can take before a collision takes place. Cf. Boltzmann, *Vorlesungen über Gastheorie*, I, §3.

23. For the correct formulation of this purely kinematic part of the problem, see Boltzmann, *Gastheorie*, I, 15, and V 8, Section 8. For the case of molecules of general structure, see Boltzmann, *Gastheorie*, I, 107; II, 230.

24. Clausius [2, 3].

25. This last assumption about equal frequencies is often referred to as the "hypothesis of molecular chaos." However, the same phrase is also applied to another, considerably deeper statement, which will be developed in Section 18c. The confusion of the two meanings has an important role in our discussion of the H-theorem. Hence it seemed mandatory to reserve the expression the "hypothesis of molecular chaos" in our discussion exclusively for the concept introduced in Section 18c. Cf. note 161.

26. Clausius [2] (1859).

27. *Ges. Abh.*, III, 12 ff.

28. As for instance in the case of the equilibrium between liquid and vapor.

29. Clausius [2].

30. Cf. note 33.

31. Maxwell [1].

32. Maxwell in his first paper also considers the case of non-spherical, rigid molecules. For the frequency of the different values which the velocity of rotation can have, he makes without further explanation an assumption analogous to (1).

33. Cf. V 8, Section 8.

34. Cf. Eq. B′, in note 179.

35. Boltzmann [2, 3].

36. E.g., in the case of gravity the vertical direction is preferred, and thus the different directions of velocity cannot be considered *gleichberechtigt*. (Compare with note 40.)

37. Boltzmann [3]. Boltzmann was led to this generalized formulation of the problem by some attempts he had undertaken (1866) [1] to derive from kinetic concepts the Carnot-Clausius theorem about the limited convertibility of heat into work. In order to carry through such a derivation for an arbitrary thermal system (Boltzmann [5], (1871)) it was necessary to calculate, e.g., for a nonideal gas, how in an infinitely slow change of the state of the system the added amount of heat is divided between the translational and internal kinetic energy and the various forms of potential energy of the gas molecule. It is just for this that the distribution law introduced above is needed.

38. For a closer definition of these limits by which (3) assumes a really precise meaning, see note 120.

39. We restrict ourselves to the case when the mutual potential energy between two different molecules can be neglected.

40. Statement (3″) contains among other things the following consequences:

a) In all parts of the container we find the same distribution of speed and hence also the same average kinetic energy (temperature).
b) All directions of velocity occur equally often, although—cf. note 36—the vertical direction is now preferred.
c) The barometric formula for the density dependence on altitude. Loschmidt [1] attacked especially statement (a). He tried to prove that in a state of equilibrium the gas is hot at the bottom and cold at the top. Cf. Boltzmann's reply [8, 11].

41. See, however, the postulate of Maxwell [2] about the anisotropic distribution of velocities in gases having a stationary shearing motion, which he states without any attempt at proof:

$$f = AC^{-(\cdots)-(B_1u^2+B_2v^2+B_3w^2)}.$$

42. Maxwell [1]. He discards this proof in his publication [2].

43. Maxwell [2].

44. Cf. V 8, Section 6.

45. Every collision changes the initial states of two molecules into new ones. Maxwell shows that, for the Maxwell distribution in each interval of time and in each interval of the velocity distribution, just as many molecules leave the velocity region as enter it.

46. Boltzmann [2, 3].

47. Boltzmann [6, 7]. Concerning the H-theorem itself, see Sections 6 and 14. The presentation in V 8 Section 11 differs significantly in the formulation of the assumptions and of the results from the original papers [6, 7].

48. To be sure, Maxwell [2] (1866) had already outlined a proof for the statement that the Maxwell distribution is the only one that will stay stationary in time. Boltzmann [6], however, showed (at the beginning of Chapter I) that this proof is faulty because of an oversight. Much later (1887) Boltzmann succeeded (cf. *Gastheorie*, II, §93) in carrying through consistently the Maxwellian method of proof. However, the approach through the H-theorem is superior because it also includes the time development of the Maxwell distribution.

49. The hypothetical character of the *Stosszahlansatz* was not realized for a long time. For evidence, see in Boltzmann [4] (1871) the final remark: " . . . and so in that paper" (referring to publication [3], which was based on this postulate) "I followed the more tedious method, which, however, is free from hypotheses."

50. Boltzmann [6, 7, 16]; Lorentz [1] (1096).

51. To be sure, Clausius and Maxwell had given, much earlier, a kinetic interpretation of viscosity, heat conduction, and diffusion. They, however, confined themselves to stationary processes, and hence the paradox we are discussing did not arise.

52. Cf. with this postulate Section 17.

53. Boltzmann in his first papers on the H-theorem still denotes this function by E (Entropy).

54. For the exact definition of $\Delta\tau$, which gives Eq. (10) a precise meaning, see note 120.

55. $\Delta\tau$ is usually written as a differential and therefore the sum as a multiple integral. For this, see Boltzmann [6, Chap. II; 10, Chap. II] and also Section 12e and note 119.

56. Cf. V 8, Section 11. In the example given in the Appendix to Section 5 H stays constant as soon as $f_1=f_2=f_3=f_4$.

57. In thermodynamics the entropy is defined only for states of equilibrium. Indeed, Boltzmann by actually computing the H-function has shown for a very general class of gas models ([6, Chap. VI; 10, Chap. V], also *Gastheorie*, I, 139) that this function, is the same, apart from an additive constant, as the negative entropy if we consider states of equilibrium. For states of nonequilibrium, $-H$ is a generalization of the thermodynamical entropy. For the combinatorial meaning of the quantity H, see Section 12d.

58. Loschmidt [1].

59. To see this, one has to go back to the strict definition of $\Delta\tau$. It is easy to verify for example (3′), and especially for the example in the Appendix of Section 5. In that case,

$$H=f_1 \log f_1+ \cdots +f_4 \log f_4;$$

furthermore $f_1'=f_3, f_2'=f_4, f_3'=f_1, f_4'=f_2$, so that also $H'=H$.

60. If in the example in the Appendix to Section 5 we invert all the velocities of the P-molecules at time t_n, the equalized velocity distribution would be reconverted into a more unequalized one.

61. Zermelo [1] (1896). Recently this *Wiederkehreinwand* has often been referred to Gibbs's *Elementary Principles in Statistical Mechanics,* Chap. XII, where Zermelo is not mentioned.

62. Poincaré [1]. A detailed discussion of the conditions for the validity of the theorem can be found in Boltzmann [20].

63. In general, among all the motions of the system we can also find some whose asymptotic development in time cannot be characterized in the above-described way. The measure of their set, however, is at least one order smaller than the measure of the set of all motions. For this reason such exceptional cases do not play any role in the present controversy. Cf. Boltzmann [20]; Gibbs, *Statist. Mech.*, Chap. XII.

64. For an attempt to give a numerical estimate, see Boltzmann [18, 19].

65. Cf., e.g., Burbury, *Treatise* (1899), §39; Lippmann [1] (1900); Liénard [1] (1903). The criticism of the *Stosszahlansatz* which is discussed there can be explained on the model outlined in the Appendix to Section 5 in the following way: Let us consider a long, undisturbed motion, and let us pick out a certain time interval Δt within the duration of this motion. Let us consider the completely reversed motion during this interval. For the direct motion, let us denote by f_1, f_2, f_3, f_4 the velocity distribution before Δt, and by $\bar{f}_1, \bar{f}_2, \bar{f}_3, \bar{f}_4$ the velocity distribution after Δt. Let us denote by $N_{12}\Delta t$ the number of (1, 2)-collisions during Δt. Let us use the following notation for the corresponding reversed motion: f'_1, f'_2, f'_3, f'_4 for the velocity distribution before Δt, and $N'_{21}\Delta t$ for the number of (2, 1) collisions during Δt. Since, apart from the direction of motion of the P-molecules, the two motions are identical, we have

(a) $$f'_1 = \bar{f}_3,\ f'_2 = \bar{f}_4,\ f'_3 = \bar{f}_1,\ f_4 = \bar{f}_2$$

and

(b) $$N_{12}\Delta t = N_{12}\Delta t.$$

If now one assumed the *Stosszahlansatz* for both the direct and the inverse motion, Eq. (b) could be transformed into (cf. Eqs. 7 and 8)

(c) $$f'_2 \cdot k \cdot \Delta t = f_1 \cdot k \cdot \Delta t,$$

from which it would follow on account of (a) that

(d) $$\bar{f}_4 = f_1.$$

This equality, however, is obviously not satisfied if in the time interval we are considering the velocity distribution $f_1 \neq f_2 \neq f_3 \neq f_4$ is still strongly unequalized. It is therefore impossible in the case of an unequalized velocity distribution for the collisions to take place during a time interval (which is part of a longer, undisturbed motion) in accordance with *Stosszahlansatz* for both the direct and the reversed motions. The number of collisions must be different from that predicted by the *Stosszahlansatz* for at least one of the motions.

66. Cf. V 8, Section 26, or Boltzmann, *Gastheorie*, II, 62. The more general case of molecules of different kinds will not be considered.

67. p_s^k is defined by $p_s^k = \partial L / \partial \dot{q}_s^k$, so that with Cartesian coordinates $p = m\dot{x}$. Cf. IV 1 (A. Voss).

68. Boltzmann, *Gastheorie*, II, §§44, 45.

69. Cf. IV 1 (A. Voss).

70. Let us take a cylinder filled with gas and a piston holding the gas in the cylinder, and let us consider the action of some continuous field of force on this system. As long as the piston is not moved and the external field of force is unchanged, the above condition is fulfilled. However, if we move the piston, for example (considering the action of the piston as a field of elastic forces), the condition is violated; in this case we act on the gas, inasmuch as we change in time the existing external field of force. Cf. Section 23b.

71. If we divide, for instance, the first $(2rN-1)$ equations of Eq. (22) by the last one, we obtain $2rN-1$ differential equations which do not contain t or dt, which in turn give us the same number of integrals of the form (23a) and (23b) which do not contain time. Finally t can be determined by a quadrature from the last equation of Eq. (22), which then leads to Eq. (23c).

72. Cf. note 118 about the relationship of the Γ(gas-phase)-space to the μ(molecule-phase)-space.

73. We will denote by G the moving image point corresponding to the phase changes of a gas model. On the other hand, Γ will denote a fixed point of a certain fixed set of (q, p) values.

74. Eqs. (22) ascribe to each point Γ a certain direction, in which it will be traversed by the G-points. If the external field of force does not depend explicitly on t (and this is the case we are considering), then these directions of propagation in the Γ-space are the same for all times. Therefore they can be used to make up $\infty^{(2rN-1)}$ fixed G-paths. Because of the single-valuedness of the dynamical equations (disregarding singular points) each Γ-point is traversed by only one G-path.

75. If for two different motions the quantities $c_1, \cdots, c_{2rN-1}$ have the same values and the two c_{2rN-s} differ by an amount Δc_{2rN}, then the two corresponding G-points traverse the same G-path, but with a constant time interval Δc_{2rN} between them. The time average of a certain function $\psi(q, p)$ of the phase, taken between $t = -\infty$ and $t = +\infty$, will therefore be the same for the two motions. Cf. note 96.

76. For a more detailed discussion of "volume," "area," etc., in Γ-space, see Lorentz [1].

77. For a three-dimensional space we would have

$$[A, E]^{(-1)} = \frac{1}{\cos(N, z)} \iint dx dy;$$

$$\cos(N, z) = \frac{\dfrac{\partial E}{\partial z}}{\sqrt{\left(\dfrac{\partial E}{\partial x}\right)^2 + \left(\dfrac{\partial E}{\partial y}\right)^2 + \left(\dfrac{\partial E}{\partial z}\right)^2}}.$$

78. The domain of the integration $\overline{(A, E)}^{(-1)}$ can be considered the projection of the region $(A, E)^{(-1)}$ on the flat space $S_{2rN} = 0$.

79. Here, contrary to Eq. (27) it is not necessary to restrict ourselves to infinitesimal regions.

80. Proof: It follows from Eq. (22) that

$$\text{(A)} \qquad \frac{\partial \dot{q}_s^k}{\partial q_s^k} + \frac{\partial \dot{p}_s^k}{\partial p_s^k} = 0.$$

Summing over all the (rN) degrees of freedom, we obtain an equation (B) for the flux of the G-points. This equation is analogous to

$$\frac{\partial u}{\partial x} + \frac{\partial v}{\partial y} + \frac{\partial w}{\partial z} = 0$$

i.e., to the condition of incompressibility in the Euler form. We can easily transform this into the Lagrangian form of the condition of incompressibility:

$$\text{(B)} \qquad \frac{\partial(\overset{1}{q_1}, \cdots, \overset{N}{p_r})}{\partial(q_{10}^1, \cdots, p_{r0}^N)} = 1,$$

which then proves the statement as given above. See Boltzmann, *Gastheorie*, I, §26; Gibbs, *Statist. Mech.*, Chap. I. Equation (A) appears first in Liouville, *J. de Math.*, 3 (1838), 348. It was used by Jacobi (*Vorlesungen über Dynamik*, p. 93, theorem of the last multiplier). The interpretation as a condition of incompressibility for the streaming in the Γ-space and its use in the theory of gases is due to Boltzmann [2, 3].

81. A graphical illustration of this theorem for systems with one degree of freedom (free fall and motion of a pendulum) can be found in G. H. Bryan, *Phil. Mag.*, (5) 37 (1895), 532. If we used as Γ-space a $(q, \dot{q})$ space instead of the (q, p) space, then in such a space the G-points would form at consecutive times regions for which the integral

$$\frac{\partial(\overset{1}{p_1}, \cdots, \overset{N}{p_r})}{\partial(\dot{\overset{1}{q_1}}, \cdots, \dot{\overset{N}{q_r}})} \int \cdots \int d\overset{1}{q_1} \cdots d\overset{N}{q_r}\, d\dot{\overset{1}{q_1}} \cdots d\dot{\overset{N}{q_r}}$$

would always give the same value. Since the Jacobian is in general a function of $\overset{1}{q_1}, \cdots, \overset{N}{q_r}$, which changes along the G-path, the G-points would in general therefore occupy regions of changing volumes at consecutive times. For the description of the streaming in phase space, the (q, p) space is therefore distinguished for the simplicity of the description. Cf. note 173.

82. Let us consider two energy surfaces, $E(q, p) = c_1$ and $E(q, p) = c_1 + \delta c_1$, which are infinitesimally close to each other. Their distance δN along the G-path is inversely proportional to Q. Therefore, in order not to change the volume $(A, E)^{(-1)}\delta N$ of the box-shaped region built on $(A, E)^{(-1)}$, the $(2rN-1)$ dimensional region $(A, E)^{(-1)}$ must contract and dilate according to Eq. (27).

83. These stationary density distributions enjoy a special position in probability assumptions. Cf. note 173(d).

84. The infinitely many G-points represent infinitely many identical copies of our gas model, which started at time t from all the possible phases and which then move independently of each other, under similar conditions (i.e., the function $E(q, p)$ is the same for all of them). The fiction of such a host of infinitely many identical and independent gas models allows us to replace certain "probability assumptions" by statistical statements. They were explicitly formulated and used first by Maxwell [3] (1878), and on this occasion he used the word "statistico-mechanical" to describe the study of such ensembles of gas models (cf. note 3). However, seven years earlier Boltzmann [4] (1871) had already worked with essentially the same kind of ensembles. (cf. note 107).

85. We will therefore look for such a distribution of the infinitely many gas samples over the different phase regions that during dt as many gas models enter each phase region as leave it.

86. $E, \phi_2, \cdots, \phi_{2rN-1}$ are $2rN-1$ mutually independent functions of the $2rN$ parameters (q, p), which remain constant along each single G-path. All other functions satisfying the same conditions can be expressed in terms of these $(2rN-1)$ functions.

87. According to Eq. (27) the volume of an infinitesimal $(A, E)^{(-1)}$ is inversely proportional to Q, while its points stream over the E-surface. Therefore the "surface density" σ changes in such a way that σQ stays constant. In order to ensure therefore the stationary behavior of $\sigma_0(q, p)$ the quantity $\sigma_0 Q$ must have the same value for all points of a certain G-path. From one G-path to any other it can still vary in an arbitrary way.

88. One should be careful to distinguish between the following two concepts: (a) an ergodic system, (b) ergodic density distribution in the Γ-space. For the relationship of (a) to (b), see note 101.

89. Boltzmann [4, Chap. II] (1871), also [15] (1887), gives among others the example of the motion of a point mass in a plane under the influence of an attractive force given by the potential $\frac{1}{2}(ax^2 + by^2)$, where the ratio a/b is irrational. This gives the open Lissajous figures when the ratio of the periods is irrational. The point mass during its motion passes arbitrarily closely to each point in a certain rectangle.

90. The Lissajous figure in the above example is in a way the projection of the corresponding G-path in the (x, y, u, v) space onto the (x, y) plane. For this reason it crosses itself; something which the corresponding G-paths never do. The G-path can be visualized topologically by an open geodetic line on the surface of a torus. This, without crossing itself, approaches arbitrarily closely all the ∞^2 points (x, y, u, v) which satisfy the equations

$$\frac{m}{2} u^2 + \frac{ax^2}{2} = c_1, \qquad \frac{m}{2} v^2 + \frac{by^2}{2} = c_2$$

(c_1 and c_2 are the energies given to the two components of the oscillation at time t_0). It is not difficult either to define purely geometrically sets of curves so that each single curve of the set approaches arbitrarily closely each of the ∞^3 points inside the torus.

91. Boltzmann [4] (1871). Cf. also end of paper [2] (1868).

92. Maxwell [3] (1878).

93. ἔργον = energy; ὁδὸς = the path: the G-path traverses all points of the energy surface. This terminology was first used by Boltzmann in [15] (1887). Maxwell and with him the other British authors use in this connection the phrase, "assumption of the continuity of path" (meaning by "path" the G-path).

94. The reasoning is about as follows: The single G-path traverses each point in phase space which is on the energy surface. On the other hand, each phase point Γ lies on only one G-path (note 74), which then leads us to statement 1. Compare esp. Boltzmann [15].

95. In this respect see the statements of Boltzmann [15] about

the phase difference between the two components of the Lissajous motion when the ratio of the periods is irrational.

96. This time average is defined by

$$\lim_{\substack{T_1=-\infty \\ T_2=+\infty}} \frac{1}{T_2 - T_1} \int_{T_1}^{T_2} \phi(q, p)dt$$

Cf. note 75.

97. In order to make the G-path of the example in note 89 approach arbitrarily closely each of the ∞^3 Γ-points, which satisfy the *one* equation:

$$\frac{m}{2}(u^2 + v^2) + \frac{ax^2 + by^2}{2} = E_0,$$

Boltzmann introduces in the (x, y) plane an infinitesimal small elastic obstacle, with which the oscillating particle collides again and again. Similar procedures are used in other examples. In this respect see also Lord Rayleigh [2]. It is on account of the complexity of the collisions of the molecules with each other and with the rough though perfectly elastic wall of the container that Boltzmann and Maxwell feel justified to assume that gas models are ergodic.

98. To elucidate the difference between the following two requirements: (I) "to approach arbitrarily closely each point of the energy surface" and (II) "to traverse each point of the energy surface," let us consider again a geodesic line of a torus for which the ratio of the two numbers of turnings in the two directions is irrational, e.g., a bit bigger that $\frac{1}{2}$. Such a geodesic will intersect the meridian at infinitely many points P_h, which are densely distributed everywhere over the circumference. On the other hand, one can state that no matter how often one turns around the torus going along the geodesic line, one will never get from a point P_h to the diametrically opposite point Q on the meridian. Because if we did, then twice as many revolutions would bring us back to P_h, contrary to our original assumption. From this one can easily see that the set of those points P_h which can be reached by a given geodesic line form a denumerable subset in the continuum of all those points on the circumference which the geodesic line approaches arbitrarily closely.

99. If, for the time being, we call systems satisfying requirement (I) of note 98 "quasi-ergodic," then, instead of statements 1), 2), and 3) in the text, we must say that for a "quasi-ergodic" system on each surface $E(q, p) = E_0$ there will be a continuum of $\infty^{(2rN-2)}$ different G-paths with different values of the constants $c_2, \cdots,$

c_{2rN-1}. Hence one cannot extend the Boltzmann-Maxwell justification of statement 3) to quasi-ergodic systems.

100. Cf. note 173(f).

101. For an ergodic system the ergodic density distribution as given in Eq. (31) would be the only one that was stationary. This follows from the fact that, on the one hand according to 94, all points of the surface $E(q, p) = E_0$ should lie on one single G-path of the system. On the other hand, according to Eq. (29), the product σQ should be constant along a single G-path. For a nonergodic system the distributions (30) and (31) are still stationary, but they appear in this case arbitrarily specialized.

102. If we restrict ourselves to continuous density distribution, $\sigma(q, p)$, then even for quasi-ergodic systems distribution (32) would be the only one which would remain stationary. We see, however, from Eq. 33, that we do not gain anything of physical value from this analytical trick. The time average in question can change quite discontinuously from path to path for a quasi-ergodic system, because we obtain it by averaging over an infinite time interval.

103. Boltzmann himself, on the bottom of p. 92 in *Gastheorie*, II, introduces distribution (32) as the "simplest case" of a stationary distribution and calls it "ergodic" without referring to the ergodic hypothesis anywhere in that textbook.

104. Boltzmann [2] (1868).

105. See the different statement to which the continuation of this investigation in Sections 13 and 14 will lead us.

106. Cf. V 8, Section 28; *Gastheorie*, II, §34.

107. Cf. Boltzmann [4, Chap. II between Eqs. 22 and 23].

108. This statement is still independent of the ergodic hypothesis.

109. For dS we refer to Eq. (25).

110. Cf. note 108.

111. Cf. Lord Rayleigh [2], Bryan [1], Lord Kelvin, *Baltimore Lectures on Molecular Dynamics* (London, 1904), Jeans, *Dynamical Theory of Gases*, §§92–95.

112. Maxwell [3]. For this see the review by Boltzmann [13].

113. Lord Rayleigh [2].

114. Jeans, *loc. cit.*

115. Boltzmann [4, Chap. III].

116. *Ibid.*, between Eqs. 22 and 23.

117. The modification of the ergodic hypothesis developed by Jeans (*Dynamical Theory*, §96 ff.) is based on the *Stosszahlansatz*. Hence this approach does not seem to be suited for a critical foundation either.

118. The direction in which a point $m^{(k)}$ will traverse a fixed μ-point will therefore depend on the simultaneous positions of all other points m and therefore is different for different times—contrary to the situation in the Γ-space as outlined in note 74.

119. A compromise between the following two requirements is necessary: (a) ω must be very small compared to the smallest physically distinguishable dimensions, (b) ω must be large enough so that the numbers a_i which we will introduce below will in general be very large. In the discussion of many questions one has to remember that one cannot go to the limit 0 for the size of the parallelopipeds ω_i. Cf. note 128.

120. The ω_i's just defined are the $\Delta\tau$'s we were talking about in notes 38 and 54.

121. When later we replace $a_i!$ by the Stirling approximation, we will be following the generally accepted procedure of Boltzmann. Requirement (b) of note 119 will be especially important in that case.

122. See for example the stationary state distributions in Eqs. (3) and (3′) which correspond to thermal equilibrium.

123. Boltzmann [10]; Jeans, *Dynamical Theory*, §§39 ff.

124. The set of numbers a_i remains completely unchanged if we subject the phase of the gas model to the following two types of transformations: (a) Transform one of the image points $m^{(k)}$ successively into all of the points μ of the cell ω_i which contains this $m^{(k)}$. All a_i's remain unchanged. They also remain so when we do the same operation simultaneously with all of the N-image points $m^{(k)}$. During this operation the G-point in Γ-space covers a certain $2rN$ dimensional "Ω cell" whose volume is

$$[\Omega] = [\omega]^N. \tag{35a}$$

(b) Let us interchange the image point $m^{(k)}$ with another $m^{(h)}$ in the μ-space. All a_i's remain the same. Hence they also remain the same if one performs in succession all the $N!$ permutations of this sort. Correspondingly the Γ-point of the gas models will assume altogether $N!$ different positions. We might perhaps call their distribution star-shaped, since they can be transformed into each other by certain finite rotations of the Γ-space about the origin.

125. Let us start from a certain phase Γ_0 and let us consider first all those permutations under which not one of the image points $m^{(k)}$ leaves the cell ω in which it was contained originally. One obtains in this way $a_1!\, a_2!\, a_3! \cdots$ Γ-points. They all lie in the same Ω-cell which we have already generated from Γ_0 by performing the operations under (a) in note 124. In order to reach a Γ-point

which belongs to another Ω-cell, we have to permute two image points $m^{(k)}$ and $m^{(h)}$ lying in different ω-cells. From these remarks we immediately see that (1) the $N!$ points, which will be generated from Γ_0 by all the permutations, can be divided into groups each containing $a_1! \, a_2! \, a_3! \cdots$ points, which lie all in the same Ω-cell, (2) that for this reason the combination of operations of (a) with the permutations will give altogether only $N!/(a_1! \, a_2! \cdots)$ different Ω cells, which proves Eq. (35).

126. See especially Boltzmann [10, Chap. V], also V 8, Section 13.

127. This is so quite independently of whether the change of the integers a_i is brought about by a collision or by collision-free motion of the molecules.

128. Most calculations require Δt to be so large that it will contain a large number of collisions of a certain kind (cf. note 119).

129. Cf. Section 14b.

130. Boltzmann [10]; Jeans, *Dynamical Theory*, §§41 ff.

131. The calculations in this case are clearly analogous to those required to prove the Bernoulli theorem. In order to show the first part of the statement, all we have to do is to determine the maximum of Eq. (36), i.e., the minimum of Eq. (43), given the auxiliary condition of Eq. (45). Boltzmann makes use of the second half of the statement in all those cases when he calls the Maxwell velocity distribution "overwhelmingly the most probable one." A more quantitative formulation and derivation of this part of the statement is sketched by Jeans in [2, §§22–26] and in *Dynamical Theory*, §§44–46 and 56.

132. Cf. Eq. (25).

133. An actual proof of statement II or its derivation from statement I does not exist. It would serve as a basis for the study of deviations from the "most probable state" (cf. Section 25). In this sense the papers quoted there give the most thorough discussion of this question.

134. Cf. Section 12e.

135. Boltzmann [17, 21]; Jeans [2, §32]; P. and T. Ehrenfest [2].

136. For a numerical treatment of the relative frequency of occurrences for a similar "curve" defined by a lottery scheme, see Ehrenfest [2].

137. At this point we should note the following: Since the H-curve attains a certain large height much more often than an even larger height, we can see that H_a also lies much more often at a maximum than on a slope leading to or from an even higher H_b.

138. The smoothened interpolating curve has only a finite num-

ber of maxima within a finite interval, while the rest of the continuum of its points forms the slopes of these maxima. Statements (Va) and (Vb) are indeed inapplicable to these interpolating curves.

139. Cf. also note 152.

140. This is again a synthesis of a collection of remarks which are dispersed over all the quoted works of Boltzmann. They show clearly how many purely intuitive statements are hidden behind the usual probabilistic terminology.

141. The measure of the dispersion should be based on the Γ-volumes of the corresponding initial states. Cf. note 173.

142. The single H-curves almost always run near H_0. Therefore it is very possible that at any time $t_A + n\Delta t$ (when n is not too small) almost all H-curves will lie near H_0 and only a vanishing fraction of them will be traversing a hump.

143. Cf. notes 141 and 173.

144. Cf. note 63.

145. Cf. in any case Section 18.

146. Cf. notes 141 and 173.

147. Cf. Section 25.

148. Cf. note 11.

149. For the last word of Boltzmann on this question, see V 8, Section 14.

150. Zermelo [2, "Theorem II"].

151. See Eq. (13).

152. This erroneous statement comes about by confusing the discrete H-curve with its smoothened interpolating curve. For the latter the left and right derivatives would be equal at every point, which leads immediately to an equation similar to Eq. (51). However, such a conclusion cannot be transferred to difference quotients, which correspond to Eq. (51) itself.

153. Cf. note 57.

154. V 8, Section 14; Boltzmann, *Gastheorie*, II, §§88, 89.

155. It is worth remarking in this connection that the Brownian motion is much more compatible with kinetic ideas than with the dogmatic formulation of the second law.

156. Boltzmann [3, end] and "Über die Natur des Gasmoleküle," *Wien Ber.*, II, 74 (1876), p. 555.

157. Cf. note 65.

158. Boltzmann [9, Chap. II; 17, 18, 19, 21]. For the model in the Appendix of Section 5 these statements can be proved easily.

159. Boltzmann, *Gastheorie*, I, 43.

160. Jeans [2] and *Dynamical Theory*, §62 ff.

161. The usual terminology unfortunately does not differentiate

between the *Stosszahlansatz* and the hypothesis of molecular chaos. See, e.g., Boltzmann, *Gastheorie*, I, 21. Similarly the French authors use the term *mouvement inorganisé* in both meanings. See also "assumption A" in the work of Burbury. Jeans clearly distinguishes the two concepts, but the phrase "molecular chaos" is still used sometimes to denote the *Stosszahlansatz.*

162. Burbury, *Treatise* (1899), §§39, 69; also in several papers in the *Phil. Mag.*, 1900–1908.

163. Because of the assumptions restricting the smallness of the cells ω_i (note 119), it depends on the precise position of the image points $m^{(k)}$ inside their ω-cells whether two molecules belonging to different cells will actually collide in the subsequent time interval Δt.

164. Jeans, *Dynamical Theory*, §§65, 66.

165. In Section 12b we neglected the corrections which are to be applied to Eq. (35a) (note 124) because of the impenetrability of molecules. Here, however, these corrections have to be taken into account. See Jeans, *loc. cit.*

166. Whether or not this statement is compatible with all the previous statements can apparently be decided only by computing the motions of all the members of the set from t_A to t_B. This would determine the composition of the subset but is clearly practically impossible.

167. See note 162.

168. Burbury has not agreed with this opinion.

169. See, e.g., the hypothesis of Krönig in Section 2.

170. Cf. J. von Kries, *Die Prinzipien der Wahrscheinlichkeitsrechnung* (Freiburg, 1886), Chap. VIII.

171. Cf. H. Bruns, *Wahrscheinlichkeitsrechnung und Kollektivmasslehre* (Leipzig, 1906), §§63, 64.

172. *Ibid.*, p. 93 ff.

173. Wherever in statements (IV)–(XI) of Sections 14–17 the phrase "overwhelming majority" was used, one should, following Boltzmann, take as a measure for the frequency of occurrence the "volumes" of the corresponding regions in Γ-space. It depends on the selection of this measure which occurrence in the group of motions will be "overwhelmingly the most frequent." One encounters here a question which arises in an analogous way in every investigation concerning "geometrical probabilities": What right did Boltzmann have to prefer this particular measure of frequency over others? Let us consider two groups of motions, and let us denote the regions which their G-points occupy in Γ-space at the consecutive times $t_1, t_2, t_3, \cdots$, by $(A_1), (A_2), (A_3), \cdots$, and $(B_1), (B_2), (B_3), \cdots$,

respectively. If we want to define a measure for the relative frequency of the motions (i.e., of their G-points) in the two groups, we can clearly admit only those measures which give the same value for the relative frequency of these G-points for all times $t_1, t_2, t_3, \cdots$; we furthermore must demand that this requirement hold for any arbitrary choice of the two initial regions (A_1) and (B_1). Let us denote this relative frequency by $\{A_s\}$: $\{B_s\}$; then our requirement is

$$\text{(I)} \qquad \frac{\{A_1\}}{\{B_1\}} = \frac{\{A_2\}}{\{B_2\}} = \cdots = \frac{\{A_s\}}{\{B_s\}}.$$

This leads immediately to the following remarks:

a) Because of Liouville's theorem (Eq. 26), Boltzmann's frequency measure $\{A_s\} = \{A_s\}$, $\{B_s\} = \{B_s\}$ indeed satisfies the invariance requirement (I).

b) If we take for the Γ-space a $(q, \dot{q})$ space instead of the (q, p) space and use for the frequency measure the volumes which the G-points occupy in this space, we will in general be in conflict with requirement (I). Cf. note 81.

c) If we take for the frequency measure the integral

$$\text{(II)} \qquad \int \cdots \int F(E, \phi_2, \cdots, \phi_{2rN-1}) dq_1^1 \cdots dp_r^N$$

taken over the region which is occupied by the G-points in the (q, p) space at any time (the meaning of F being the same as in Eq. (28)), then this definition of the measure satisfies requirement (I) for any arbitrary F function provided that the F is chosen once and for all.

d) The definition of the class of measures given in (c) uses for weight function the most general stationary density distribution (Eq. 28) in Γ-space.

e) A weight function which cannot be expressed in the form $F(E, \phi_2, \cdots, \phi_{2rN-1})$ cannot simultaneously satisfy requirement (I) and the requirement that it give the same measure for a given volume element in Γ-space at all times t.

f) The restriction of the weight function to the much more special form $F(E)$ is equivalent to the restriction to the "ergodic" density distribution as given in Eq. (30) in Γ-space. (Since Sections 13–15 always deal only with a group of motions with the same E, there the Boltzmann measure $F=1$ does not give results different from the most general ergodic measure $F=F(E)$).

Summarizing we can say: Boltzmann's definition of measure is not quite arbitrary in the sense that Liouville's theorem together with requirement (I) excludes all measures except the class defined in (c) of which Boltzmann's measure is the simplest special case. The restriction of the weight function $F(E, \phi_2, \cdots, \phi_{2rN-1})$ to $F(E)$ is arbitrary, however, if one disregards the assertion that gas models are ergodic systems. The fact that this specialization is believed necessary also in those cases where it is obviously inexpedient (cf. end of Section 26) can be partly explained on historical grounds. The pioneering works of Boltzmann and Maxwell that founded the methods of statistical mechanics were completely dominated by this ergodic hypothesis.

174. See J. Hadamard [1] and H. Poincaré [4].

175. Gibbs's discussion (*op. cit.*, p. 171) shows that in the following he sticks little closer to such an axiomatization program than, say, Maxwell and Boltzmann.

176. See also the ensembles referred to in note 194.

177. The remarks following Eqs. (61) and (62), together with the considerations at the beginning of Section 24a and the middle of Section 25, try to illuminate the reasons which might have prompted Gibbs to introduce this special ensemble distribution. See Gibbs, *op. cit.*, Chap. II and Chap. XIV, pp. 187–188.

178. In the example (note 70) such parameters are the intensity of the force field and the position of the piston.

179. *Op. cit.*, Chap. VII, p. 73. See the exception discussed in note 2, p. 75. The relationship of the smallness of the dispersion with the very large value of $2rN$ in statement (XII) can be understood by considering the following example (*ibid.*, p. 79): Let us assume that the gas quantum consists of N-point molecules which are elastically bound to fixed positions of equilibrium. The co-ordinates x, y, z of each molecule will be counted from its position of equilibrium. Then $q_1 = x_1,\ q_2 = y_1,\ \cdots,\ q_{3N} = z_N,\ p_1 = m\dot{x}_1,\ \cdots,\ p_{3N} = m\dot{z}_N$; therefore

$$E = \frac{x}{2} \sum_{h=1}^{3N} q_h^2 + \frac{1}{2m} \sum_{h=1}^{3N} p_h^2.$$

The group of surfaces $E = \text{const.}$ consists of a set of similar "ellipsoids" whose surface increases very rapidly with E, namely as $(\sqrt{E})^{6N-1}$. The number of individuals in ensemble (53) for which $\sqrt{E}$ lies in the infinitesimal range $d(\sqrt{E})$ is given by

$$\text{(A)} \qquad \text{const. } R^{6N-1} \cdot c^{-R^2/\Theta^2} dR$$

(where R is used as an abbreviation for $\sqrt{E}$). The most frequent value, the mean, and the mean square of the deviation from the average are

$$E_0 = (6N - 1)\frac{\Theta}{2}, \qquad \bar{E} = 6N\frac{\Theta}{2}, \qquad \overline{(E - \bar{E})^2} = 3N\Theta^2$$

respectively. Using the usual measure we get

$$\text{(B)} \qquad \frac{\overline{(E - \bar{E})^2}}{(\bar{E})^2} = \frac{1}{3N},$$

which is the required "dispersion." One should compare this distribution (A) of the ensemble of gas models with the Maxwell distribution of speeds c in an ensemble of point molecules and with the dispersion which it gives for the kinetic energy λ of these molecules:

$$\text{(A}'\text{)} \qquad \text{const.}\ c^2 \cdot e^{-\alpha c^2} dc \qquad \text{(B}'\text{)} \qquad \frac{\overline{(\lambda - \bar{\lambda})^2}}{(\bar{\lambda})^2} = \frac{1}{3}.$$

180. This term was introduced by Boltzmann (1882) [13]. At the same place we can find a discussion of the difference between the momentum and its generalization, the momentoid.

181. The thesis of L. S. Ornstein (1908) [1] contains a thorough investigation of this question. See in particular p. 119.

182. Boltzmann [4]. The transition from Eq. (57) to Eq. (57′) of Section 24a is only possible with the help of a series of additional assumptions.

183. See note 181.

184. Gibbs denotes this function as well as the function $\sum$ introduced by Eq. (66) by $\bar{\eta}$. For a proof of statements (XIII)–(XIV), see Gibbs, *op. cit.* Chap. XI.

185. Lorentz (1906) [2] §79.

186. Gibbs, *op. cit.*, p. 174: "The microcanonical ensemble of systems in which all have the same energy . . . in many cases represent simply the time ensemble, i.e., the ensemble of the phases through which a single system passes in the course of time."

187. T. and P. Ehrenfest (1906) [1].

188. Let (q, p) be the co-ordinates of a G-point at time t if its position at time 0 was (q_0, p_0). Then

$$\sigma(t) = \int_{-\infty}^{+\infty} \cdots \int \rho \log \rho \, dq_1^1 \cdots dp_r^N$$
$$= \int_{-\infty}^{+\infty} \cdots \int \rho \log \rho \frac{\partial(q_1^1, \cdots, p_r^N)}{\partial(q_{10}^1, \cdots, p_{r0}^N)} dq_{10}^1 \cdots dp_{r0}^N.$$

According to note 80 the Jacobian is 1, and by Eq. (26′) we have $\rho \log \rho = \rho_0 \log \rho_0$, which means that the integral is $\sigma(0)$. However, a statement analogous to Eq. (26′) holds for P only approximately. Cf. note 200.

189. *Op. cit.*, p. 153 at the top.

190. *Ibid.*, p. 153. This is particularly remarkable because Gibbs obviously felt that his proof gave more than it should according to the suppositions. The penultimate paragraph of Chapter XII, so important for the theory of irreversibility, is incomprehensible to us.

191. These statements appear in Chapter XIII, where they are derived, in a different form than in Chapter XIV, where they are used in the discussion of the thermodynamical analogies. We will follow the latter formulation and the presentation of Lorentz [2].

192. See in particular Gibbs, *op. cit.*, p. 165, where the proof of statement XIX ends. Cf. note 200.

193. Boltzmann [3]. There it is proved for the special case of Cartesian co-ordinates. Cf. Boltzmann, *Gastheorie*, II, 124.

194. For the case when, besides the total energy, the values of other integrals of the motion are given, such as for a free system the components of the angular momentum $F_1(q, p)$, $F_2(q, p)$, $F_3(q, p)$, Boltzmann considers the corresponding regions of Γ-space, while Gibbs works with the ensemble distribution exp $[(\psi - E)/\Theta + F_1/\Omega_1 + F_2/\Omega_2 + F_3/\Omega_3]$; (*op. cit.*, p. 37). In the combinatorial treatment which Boltzmann (1883) [14] uses to derive the distribution of states in the dissociation equilibrium, besides the total energy the numbers of atoms of various types also occur as auxiliary conditions. Gibbs with his formal trick manages to evade these auxiliary conditions also. In his treatment of the dissociation equilibrium in Chapter XV he achieves this by introducing a peculiar ensemble of gas models, in which in the different members of the ensemble the numbers of the types of atoms vary and cover the whole range of integers from zero to infinity. *Op. cit.*, p. 198.

195. Boltzmann (1894) [17a] in co-operation with G. H. Bryan has improved the completeness of his older considerations in some respects.

196. *Op. cit.*, p. 35.

197. Cf. note 192.

198. *Op. cit.*, pp. 158 ff.

199. This assumption is sufficient to allow the replacement of $\overline{\delta E}$ by $\delta\overline{E}$, as is done in Eq. (74). Identity (Eq. 73) is based on the assumption that the varied ensemble distribution is also canonical.

200. Every ρ-distribution of type (52) remains stationary and hence gives a $\sum$ which is constant in time. In this connection it may be instructive to consider the irreversible processes from Boltz-

mann's viewpoint in the idealized system of a gas containing N-point molecules which do not interact with each other and which are subject to no outside forces except elastic repulsion when they collide with the irregularly formed, but perfectly elastic, wall of the container. In this case, the N-points representing the states of the N-molecules in μ-space move as independently of each other as the ∞ many points of the ensemble move in Γ-space. An arbitrarily prescribed a_i-distribution in μ-space will in general gradually rearrange itself by an equalization of the inhomogeneities in density in the container and by the disappearance of preferred directions of motion. Therefore the statement that $\lim_{t=\infty} H(t) < H(t_0)$ is quite plausible. However, the statement that this limit coincides exactly or almost exactly with the H-value corresponding to the appropriate Maxwell-Boltzmann distribution obviously cannot be accepted. In this case one cannot expect that the Maxwell-Boltzmann distribution will approximately be reached in time since the absolute value of all the velocities remains the same throughout the process. It is worth noting that Boltzmann always neglects the changes brought about in H by the collision-free rearrangement of the a_i distribution compared to the changes which are caused by the collisions between molecules. Cf. Boltzmann (1872) [6, Chap. V] and *Gastheorie*, I, §18, note 2, and II, §75. The fact that Boltzmann and all the other authors following him obtain *exactly* the value 0 for the change of H owing to this rearrangement can be explained by their approximation of the discontinuous a_i-distribution by a continuous one. Therefore they work with an η-function which is analogous to H in the same way as in note 188 the quantity σ is analogous to $\sum$.

201. M. Planck (1904) [1].

202. L. S. Ornstein (1908) [1], pp. 57, 58.

203. H. A. Lorentz [2], §83.

204. Boltzmann [4, Chap. III].

205. Smoluchowski [1].

206. Smoluchowski [4].

207. A. Einstein [3, 4].

208. Smoluchowski [2, 3].

209. Concerning the experimental verification, cf. the work of Perrin cited in note 213.

210. Further applications to new fields can be found in the work of P. Langevin: "On the Recombination of Electrically Dissociated Gases" (Thèse, Paris 1902) and "On the Magnetic Permeability of Gases" (*J. d. Phys.*, 4 [1905], 678).

211. What chiefly matters here is that the electrons and gas ions

on account of their electric charge are much easier to handle and control than the electrically neutral molecules. Besides, under certain circumstances one single ion can produce a well-observable discharge phenomenon. How surprisingly far one can follow a single atom by exploiting this circumstance is illustrated by two new papers, by E. Regener, *Verh. d. Berlin. Phys. Ges.*, 10 (1908), 78, and by E. Rutherford and H. Geiger, *Phys. Zeitschr. Jhrg.*, 10 (1908–1909), 1. In these, 3–5 alpha particles per minute sent out by a certain amount of radioactive substance through a small metal window are counted one by one by optical or electrical discharge effect, respectively.

212. O. W. Richardson, *Phil. Mag.*, (6) 16 (1908), 353, 890; *ibid.*, 18 (1909), 681. The first paper was done with the collaboration of F. C. Brown.

213. M. von Smoluchowski (1906) [3]; J. Perrin (1910) [1]. Cf. also the papers of A. Einstein cited in Section 14d.

214. E.g., Perrin, *op. cit.*, §28 end. For the dependence of the effect on temperature, see Richardson.

215. V 23, Section 11. The fact that natural light is unpolarized and that the light rays originating from different sources of emission cannot interfere with each other, i.e., that they are incoherent, has already been explained by Fresnel by assuming that in each collision the phase and the direction of the vibration of the light emitted by the molecules change abruptly, and that all possible phases and directions are equally likely. Cf. also the application of probabilistic considerations by Lord Rayleigh: "On the Resultant of a Large Number of Vibrations of the Same Pitch and of Arbitrary Phase," *Scient. Pap.*, I, No. 6 (1871), and No. 68 (1880). Similar considerations by Van der Waals, Jr. (1900) [1, Chap. IV] and by H. A. Lorentz ("Emission u. Absorption d. Metalle," *Proc. Akad. Amsterdam* [1903], p. 666) are applied to the following problem: How much radiation is emitted on the average by a group of electrons as a result of their accidental deviation from the most probable distribution. (See also Einstein (1906) [4], §2.)

216. Cf. O. Schönrock: "Breite der Spektrallinien nach dem Dopplerschen Prinzip," *Ann. d. Phys.*, 20 (1906), 995. There one can find also a discussion of papers on this subject by F. Lippich (1870), Lord Rayleigh (1889 and 1905), C. Godfrey (1901), and A. Michelson (1902).

217. Cf. V 23 (W. Wien), Section 6.

218. V 8, Section 28.

219. In the discussions about the validity of the Rayleigh-Jeans radiation formula it has been stated repeatedly that the theorem

about the equipartition of energy is an immediate consequence of the Hamiltonian canonical equations (e.g., H. A. Lorentz at the International Mathematical Congress in Rome and V 23, Section 7). One should remark, however, that the theorem can be derived only if one combines the Hamiltonian equations with the rather dubious ergodic hypothesis or with the assumption as formulated in the text. Cf. note 173.

220. T. and P. Ehrenfest [1].

221. Jan Kroò, "Über den Fundamentalsatz der statistische Mechanik," *Ann. d. Phys.*, 34 (1911), 907.

222. The treatment of this "reduced" problem by Poincaré (1906) [4] is described by Kroò as not unobjectionable.

223. The calculations of Kroò would of course give exactly the same results for $t=-\infty$. The only distinction of $t=t_0$ from $t=\pm\infty$ is that the calculations (Fourier expansion and integration by parts) assume certain restrictions about the spatial variation of the ρ-distribution at time $t=t_0$ which are violated for $t=\pm\infty$.

224. P. Hertz, "Über die mechanischen Grundlagen der Thermodynamik," *Ann. d. Phys.* 33 (1910), 225, 537.

225. *Ibid.*

226. For the precise formulation of these assumptions we refer to §§5–8 of the original paper.

227. P. Hertz, "Über die Kanon. Gesamtheit," *Versl. Amsterdam*, 24, XII, 1910.

228. Boltzmann [4, Chap. III]. The analytical apparatus of Hertz differs from that of Boltzmann. Note especially in Hertz the transition from a "quasi-canonical" ensemble to a "canonical" ensemble, i.e., the substitution of the expression $h(E_0-E)$ by $e^{-h(E-E_0)}$ for the very small values of $(E-E_0)$, which are the only ones coming under consideration.

229. This is also referred to in Van der Waals, Jr., *Ann. d. Phys.*, 35 (1911), 185.

230. L. S. Ornstein, "Some Remarks on the Mechanical Foundation of Thermodynamics," *Proc. Amsterdam*, 28, I, 1911, and 25, II, 1911.

231. See note 173.

232. A. Einstein, *Ann. d. Phys.*, 34 (1911), 175.

233. P. Debye, *Ann. d. Phys.*, 33 (1910), 441.

234. Cf. A. Einstein [2, §7].

235. In this connection cf. also Gibbs, *Statist. Mech.*, Chap. XV.

236. *Ibid.*, Chap. XIV. In Gibbs's notation the functions $\bar{\eta}$ log V, and Φ. Concerning $\bar{\eta}$ cf. note 184.

237. In the case of an ideal gas we can verify this by a direct evaluation of the quantities $\bar{\eta}$, log V, and Φ.

238. To verify the correctness of this and the following remarks, let us consider first that measure of the entropy which is designated by Gibbs as Φ. Φ is defined by the equation (*Statist. Mech.*, Eq. 226) as

$$\Phi = \log \frac{dV}{dE}$$

where $V(E)$—see *ibid.*, Eq. 265—denotes the volume of the region in Γ-space for which the total energy of the gas is smaller than or equal to E. Let us note now that

$$V(E + dE) - V(E)$$

measures the volume of an infinitesimal "energy shell." By a simple argument (cf. notes 77 and 87) we get from this the following result: The quantity dV/dE measures the total set of G-points belonging to the ergodic surface density distribution (31) [Section 10b] on the energy surface E in Γ-space. Hence the statement in the text is indeed valid for Gibbs's measure Φ of the entropy. There then remains only the question of why, for a system with a large number of degrees of freedom, the other two measures of entropy $\bar{\eta}$ and log V very nearly coincide with Φ.

239. Cf. note 194. In this connection let us mention an interesting statistical problem which emerges if (e.g., in connection with the photochemical and thermoelectric phenomena) one poses the following question: What are the statistical and with it the thermodynamical features which distinguish the stationary irreversible processes from the nonstationary ones? That in this case also one can obtain, through a calculation used for the H-theorem, the macroscopical differential equations for the distributions of temperature, pressure, etc., by the explicit introduction of the stationarity assumption has been shown by Boltzmann (see, e.g., *Gastheorie*, I, 144, Eq. $B_1(\phi) = 0$). This remark has also been used at least partially in the kinetic theory of thermoelectricity. (See H. A. Lorentz, *Theory of Electrons* [Leipzig, 1909], p. 271.) However, the following problem, which necessarily follows from Boltzmann's train of thought, has remained untouched. Consider an irreversible process which, with fixed outside constraints, is passing by itself from the nonstationary to the stationary state. Can we characterize in any sense the resulting distribution of state as the "relatively most probable distribution," and can this be given in terms of the minimum of a function which can be regarded as the generalization

of the H-function? (Just as in the case of equilibrium, here also $dH/dt=0$. However, it is not possible to characterize the stationary state by a relative minimum of H without further specifications.) A related thermodynamical problem is discussed by W. Nernst, "Das Dissoziationsgleichgewicht im Temperaturgefälle," *Festschrift für L. Boltzmann* (Leipzig, 1904), p. 904.

240. W. Nernst, *Berl. Ber.*, 1910, p. 262; 1911, p. 306; F. A. Lindemann, *ibid.*, p. 316.

241. Cf. V 23, Section 6. In the case of sinusoidal oscillations the time average of the potential energy is equal to that of the kinetic energy. The theorem of equipartition of kinetic energy therefore determines also the total energy content for harmonic oscillators.

242. A. Einstein, "Die Plancksche Theorie der Strahlung und die Theorie der spezifischen Wärme," *Ann. d. Phys.*, 22 (1907), 180, 800. See also O. Sackur, *Ann. d. Phys.*, 34 (1911), 455.

243. If the atomic mass is small, one can expect in general a higher frequency and therefore a too small specific heat even at normal temperatures; in the case of large atomic masses this happens only at very low temperatures.

244. V 23 (W. Wien), Section 4. Planck's assumption that a resonator of frequency ν can have only energy values which are integer multiples of the elementary quantum $h\nu$ (h being a universal constant) implies that the G-point of a system of resonators with various frequencies cannot reach all the points in the Γ-space, but only the points of a certain discrete set of domains of lower dimensionality. These domains are distributed along the different co-ordinates of the Γ-space with different densities. It is the least dense in those directions which correspond to the co-ordinates and momenta of oscillators with the highest frequency. (The proof that a uniform ρ distribution of these domains of Γ-space remains stationary—the analogy to the theorems in Section 9d—is not presented by Planck because he does not discuss the mechanism which causes the oscillator to change its energy in jumps.)

245. W. Nernst and F. A. Lindemann, *Berl. Ber.*, 1911, p. 494, discuss the deviations from Einstein's result. P. Ehrenfest, "Welche Rolle spielt die Lichtquantenhypothese in der Theorie der Wärmestrahlung?" *Ann. d. Phys.*, 36 (1911), 91, studies the possibility of a generalization of Planck's assumption in the field of black-body radiation.

246. In proving that $\delta Q:T$ is a perfect differential, both Boltzmann and Gibbs always use ergodically distributed ensembles of systems. It is only in connection with his criticism of the Helmholtz

monocycle analogies (cf. Section 24e) that Boltzmann also investigates a few special examples of nonergodically distributed ensembles of systems in order to show that for these the analogies to thermodynamics generally fail. In connection with the problem of thermal radiation P. Ehrenfest (*Journ. d. russ. phys. ges.*, 43 [1911]) constructs a very general class of nonergodically distributed ensembles, for which the relation $\delta Q : T = \delta \log W$ remains valid.

247. A. Einstein, "Zum gegenwärt. Stand des Strahlungsproblems," *Phys. Zeitschr.*, 10 (1909), 195, §6.

248. Besides the papers of J. Perrin, E. Regener, E. Rutherford, and H. Geiger, quoted in Section 26, the following new investigations should be emphasized as being quite interesting in method: E. Rutherford and H. Geiger, "Probability Variations in the Emission of α Particles," *Phil. Mag.*, 20 (1910), 698; H. Geiger and E. Marsden, *Phys. Zeitschr.*, 11 (1910), 7; and T. Svedberg, "Nachweis der von der kinetischen Theorie geforderten Bewegung gelöster Moleküle," *Zeitschr. f. physik. Chemie*, 74 (1910), 738, and "Neue Methode z. Prüfung der Gültigkeit des Boyle-Gay-Lussacschen Gesetzes für kolloide Lösungen," *ibid.*, 73 (1910), 547; 77 (1911) 145. See also the papers quoted in note 250. One of Millikan: "Das Isolieren eines Ions . . . ," *Phys. Zeitschr.*, 11 (1910), 1097, contains a new method of observation that leads to a completely new counting aid in molecular statistics.

249. Besides the fields mentioned in Section 26 (notes 210, 215, 216) and the phenomena connected with radioactive decay, one should mention the following list:

Diffuse scattering of light by molecules. This comes about only because the spatial distribution of molecules has chance deviations from the regular lattice arrangement. L. Mandelstam, *Phys. Zeitschr.*, 8 (1907), 608; 9 (1908), 308, 641; M. Planck, *ibid.*, 8 (1907), 906; 9 (1908), 354; R. Gans and H. Happel, *Ann. d. Phys.*, 29 (1909), 277; H. A. Lorentz, *Proc. Amsterdam*, 25. VI, 1910; A. Einstein, *Ann. d. Phys.*, 34 (1911).

Magnetic and electric double refraction in liquids. Its explanation by the suspension of unobservable little crystallites whose complete alignment is impeded by the heat motion. A. Cotton and H. Mouton, *Bull. soc. de phys.*, 1910, p. 189; P. Langevin, *Le Radium*, 7 (1910), 249; O. M. Corbino, *Phys. Zeitschr.*, 11 (1910), 756.

Hypothesis of a spatially discrete structure of radiation fields. A. Einstein, *Ann. d. Phys.*, 17 (1905), 132; 20 (1906), 199; *Phys. Zeitschr.*, 10 (1909), 185, 817; J. Stark, *Phys. Zeitschr.*, 10 (1909), 902; 11 (1910), 25; A. Joffè, *Ann. d. Phys.*, 36 (1911); E. von

Schweidler, *Phys. Zeitschr.*, 11 (1910), 255, 614; N. Campbell, *Cambr. Phil. Soc. Proc.*, 15 (1909), 310, 513. Criticism of the hypothesis by M. Planck, *Ann. d. Phys.*, 31 (1910), 758.

Deflection of α particles in their passage through matter. H. Geiger, *Lond. Roy. Soc. Proc.* A, 81 (1908), 174; 83 (1909), 492.

We should also mention in this connection the new investigations on the thermomechanical behavior of highly rarified gases. There are a series of experimental papers by Knudsen, *Ann. d. Phys.*, 28 (1909), 75, 999; 29 (1909), 179; 31 (1910), 205, 633; 32 (1910), 809; 33 (1910), 1435; 34 (1911), 593, 823; 35 (1911), 389, and some theoretical papers by M. von Smoluchowski, *Ann. d. Phys.*, 33 (1910), 1559; 34 (1911), 182; *Phil. Mag.*, 21 (1911), 11, and by P. Debye, *Phys. Zeitschr.*, 11 (1910), 1115. Of interest here are the mechanical and thermal peculiarities which arise when the density is so low that the mean free path of a molecule is of the same order as the dimensions of the container.

250. E. von Schweidler, *Int. Congr. de Rad.*, (Liège, 1905); K. W. F. Kohlrausch, *Wien Ber.*, 115 (1906), 673 (an application to the determination of the number of α-particles emitted by 1 gram of radium per second; the direct counting succeeded only later, see note 211); a thorough discussion of all the relevant questions by N. Campbell, "The Study of Discontinuous Phenomena," *Cambr. Phil. Soc. Proc.*, 15 (1909), 117, 310, 513, also *Phys. Zeitschr.*, 11 (1910), 826; E. von Schweidler, "Zur experimentellen Entscheidung d. Frage nach d. Natur der γ Strahlen," *Phys. Zeitschr.*, 11 (1910), 225, 614; E. Meyer, "Struktur der γ Strahlen," *Berl. Ber.*, June 1910.

251. W. Lexis, *Einleit. in Theorie d. Bevölkerungs-Statistik* (Strassburg, 1875); *Massenerschein. in der menschl. Gesellschaft* (Freiburg i. Br., 1877); *Abhandlungen zur Theorie d. Bevolker.- u. Moral-Statistik* (Jena, 1903); L. Bortkiewicz, *Das Gesetz d. kleinen Zahlen* (Leipzig, 1898); "Über den Präzisionsgrad des Divergenzkoeffizienten," *Mitteil. d. Verbandes d. österr. Versich.*, Techniker Heft V, Wien, 1901; K. Pearson, "On the Criterion, That a Given System of Deviations . . . Has Arisen from Random Sampling," *Phil. Mag.*, (5) 50 (1900). Also the works of Pearson and his students in the *Journal Biometrica* (Cambridge). Cf. *idem.* 4a: L. Bortkiewicz, "Anwend. d. Wahrscheinlichkeitsrechn. auf Statistik."

252. E.g., the Galton-Pearson "Correlation coefficient"; see G. F. Lipps, "Bestimmung d. Abhängigkeit zwischen der Merkmalen eines Gegenstandes," *Ber. d. sächs. Ges. d. Wiss.*, 1905. Also the formal apparatus to describe extensive statistical ensembles, cf. H. Bruns, *Wahrsch.-R. u. Kollektivmasslehre* (Leipzig, 1906). Com-

pare also the calculations of H. Bateman, "On the Probability-Distribution of α Particles," *Phil. Mag.*, 20 (1910), 704, with the calculations of L. Bortkiewicz, *Gesetz d. kleinen Zahlen* (Leipzig, 1898).

253. The technique of such physical mass observations are discussed by E. Rutherford and H. Geiger, *Phil. Mag.*, 20 (1910), 698; T. Svedberg, *Zeitschr. f. physik. Chemie*, 77 (1911), 145.

254. The extension of the H-theorem as stimulated by the *Umkehreinwand* and *Wiederkehreinwand* (Section 14) is an example for a transition of a hypothesis of a certain order to that of the next order; the same holds for the transition from the *Stosszahlansatz* to the "hypothesis of molecular chaos" (cf. Section 18). Cf. in this connection L. Bortkiewicz, "Über den Präzisionsgrad des Divergenzkoeffizienten" (cited in note 251), where the dispersion of the dispersion is investigated. Relevant is also the sarcastic remark of Poinsot in connection with Poisson's investigations: "*Après avoir calculé la probabilité de l'erreur dans une certaine chose, il faudrait calculer le probabilité de l'erreur dans son calcul.*"

255. A formulation of such difficulties with the help of the concepts of "hypothesis of the first, second, . . . , kth order" is given by T. Ehrenfest, "Die Anwend. d. Wahrsch.-R. auf gesetzmässige Erscheinungen," *Journ. d. russ. phys. Ges.*, 43 (1911), 256 (also published in the *Phys. Zeitschr.*)

256. Cf. the statements of Boltzmann cited in Section 17.

257. M. Planck, *Acht Vorlesungen über theoret. Physik.* (Leipzig, 1909), 3rd lecture. The "special physical hypothesis" introduced by Planck to exclude the spontaneous occurrence of observable decreases in entropy (he calls it the hypothesis of "elementary disorder") consists of the following statement: The number of collisions which take place in a real gas never deviates appreciably from the *Stosszahlansatz* (cf. Section 18). The hypothesis denoted in Section 18c as the "hypothesis of molecular chaos" would, on the other hand, permit such deviations.

258. One should compare this view of Planck with that of D'Alembert, *Doutes et questions sur le calcul des probabilités* (Mélanges de litérature, d'histoire et de philosophie, vol. V, Amsterdam, 1770), with regard to the occurrence of very unlikely sequences in gambling. (Discussed by J. von Kries, *Prinzip d. Wahrsch.-R.* [Freiburg i. Br., 1886], p. 278, and H. Bruns, *Wahrsch.-R. u. Kollektivmasslehre* [Leipzig, 1906], p. 217.) The same view has often been expressed since, most recently by K. Marbe, *Naturphilosophische Untersuchungen zur Wahrscheinlichkeitslehre* (Leipzig, 1899). The treatment of Marbe was criticized by W. Lexis,

Abhandl. z. Theorie d. Bevölk.-Statistik (Jena, 1903), p. 222; L. Bortkiewicz, "Wahrsch.-Theorie u. Erfahrung," *Zeitschr. f. philos. Kritik*, 121 (1902); G. F. Lipps, "Theorie d. Kollektivgegenstände," *Philos. Studien* (Wundt), 17 (1901), 116, 575. *Ibid.*, p. 462, has the answer of Marbe.

259. In the question whether values which do not fit at all into the set of other values should be included when the average is being formed.

260. The comprehensive critical discussions of the relevant literature can be found in E. Czuber, "Die Entwickl. d. Wahrsch.-Theorie u. Ihrer Anwend." *Bericht deutsch. Mathem.-Verein.* (Leipzig, 1899); J. Kries, *Prinzip d. Wahrsch.-R.* (Frieburg i. Br., 1886); and A. Tschuprow, *Otscherki po Teorii Statistiki* (St. Petersburg, 1909), in Russian.

Bibliography

Monographs

L. Boltzmann, *Vorlesungen über Gastheorie*, 2 Bde, Leipzig 1896/1898.

S. H. Burbury, *Kinetic theory of gases*, Cambridge 1899.

J. W. Gibbs, *Elementary principles in statistical mechanics*, New York-London 1902. (German translation by E. Zermelo, Leipzig 1905.)

J. H. Jeans, *Dynamical theory of gases*, Cambridge 1904.

H. W. Watson, *Kinetic theory of gases*, Oxford 1876, 2d ed., Oxford 1893.

Papers*

L. Boltzmann, *Wissenschaftliche Abhandlungen*. Fr. Hasenöhrl, editor. Three volumes, Leipzig, 1909. (Of the following papers Nos. [1]–[6] can be found in Volume I, Nos. [7]–[13] in Volume 2, and Nos. [14]–[21] in Volume 3.)

—— [1] Mechanische Bedeutung des zweiten Hauptsatzes, *Wien Ber.* 53^2 (1866), p. 199.

—— [2] Studien über das Gleichgewicht der lebend. Kraft zwischen bewegeten materiellen Punkten, *Wien Ber.* 58^2 (1868), p. 517.

—— [3] Wärmegleichgewicht zwischen mehratomigen Gasmolekülen, *Wien Ber.* 63^2 (1871), p. 397.

* The numbers in square brackets correspond with the numbers inserted after authors' names when papers are referred to in the text.

—— [4] Einige allgemeine Sätze über Wärmegleichgewicht, *Wien Ber.* 63^2 (1871), p. 679.

—— [5] Analytischer Beweis des zweiten Hauptsatzes aus den Sätzen über Gleichgewicht der lebendigen Kraft, *Wien Ber.* 63^2 (1871), p. 712.

—— [6] Weitere Studien über Wärmegleichgewicht unter Gasmolekülen (*H*-Theorem), *Wien Ber.* 66^2 (1872), p. 275.

—— [7] Wärmegleichgewicht von Gasen, auf die äussere Kräfte wirken, *Wien Ber.* 72^2 (1875), p. 427.

—— [8] Aufstellung und Integration der Gleichungen, welche Molekularbewegung bestimmen, *Wien Ber.* 74^2 (1876), p. 503.

—— [9] Einige Probleme der mechanischen Wärmetheorie, *Wien Ber.* 75^2 (1877), p. 62.

—— [10] Beziehung zwischen dem zweiten Hauptsatz der Wärmetheorie und der Wahrscheinlichkeitsrechnung resp. den Sätzen über das Wärmegleichgewicht (Complexionen-Theorie), *Wien Ber.* 76^2 (1877), p. 373.

—— [11] Weitere Bemerkungen über einige Probleme der mechanischen Wärmetheorie, *Wien Ber.* (78) 2 (1878), p. 7.

—— [12] Einige das Wärmegleichgewicht betreffende Sätze, *Wien Ber.* 1881, p. 136.

—— [13] Referat über die Abhandlung von J. C. Maxwell [3], *Beibl. d. Ann. d. Phys.* 5 (1881), p. 403; *Phil. mag.* (5) 14 (1882), p. 299.

—— [14] Über Arbeitsquantum, welches bei chemischen Verbindungen gewonnen werden kann, *Wied. Ann.* 18 (1883), p. 309.

—— [15] Über die mechanischen Analogien des zweiten Hauptsatzes der Thermodynamik, *J. F. Math.* 100 (1887), p. 201.

—— [16] Neuer Beweis zweier Sätze über Wärmegleichgewicht unter mehratomigen Gasmolekülen, *Wien Ber.* 95^2 (1887), p. 153.

—— [17] (Three letters in *Nature* about the *Umkehreinwand*), *Nature* 51 (1894), p. 413, 581; 52 (1895), p. 221.

—— [17a] Über die mechanische Analogie des Wärmegleichgewichts zweier sich berührender Körper (jointly with G. H. Bryan), *Wien Ber.* 103^2 (1894), p. 1122.

—— [18], [19] Entgegnung auf die wärmetheoretische Betrachtung des Herrn Zermelo, *Wied. Ann.* 57 (1896), p. 773; 60 (1897), p. 392.

—— [20] Über einen mechanischen Satz von Poincaré, *Wien Ber.* 106² (1897), p. 12.

—— [21] Über die sogenannte *H*-Curve, *Math. Ann.* 50 (1898), p. 325.

E. Borel [1] Sur les principes de la théorie cinétique des gaz, *Ann. de l'école norm.* (3) 23 (1906), p. 9.

M. Brillouin [1] Introduction and notes in: Boltzmann, *Leçons sur la théorie des gaz*, trans. by A. Galotti and H. Bénard, 2 vols., Paris 1902–1905.

G. H. Bryan [1] Report on the present state of our knowledge of thermodynamics, *Report of the British Assoc.*, Cardiff 1891, Oxford 1894.

—— [2] Energy acceleration, *Arch. néérland.* (2) 5 (1900), p. 279 (also Lorentz-Jubelb.).

S. H. Burbury [1] On the law of partition of energy, *Phil. mag.* (5) 50 (1900), p. 584.

—— [2] On the variation of entropy as treated in W. Gibbs' *Statistical Mechanics*, *Phil. mag.* (6) 6 (1903), p. 251.

—— [3] On Jeans theory of gases, *Phil. mag.* (6) 6 (1903), p. 529.

—— [4] Theory of diminution of entropy, *Phil. mag.* (6) 8 (1904), p. 43.

R. Clausius [1] Über die Art von Bewegungen, die wir Wärme nennen, *Pogg. Ann.* 100 (1857), p. 253.

—— [2] Über die mittlere Länge der Wege, welche bei der Molekularbewegung gasförmiger Körper von den einzelnen Molekülen zurückgelegt werden; nebst einigen anderen Bemerkungen über die mechanische Wärmetheorie, *Pogg. Ann.* 105 (1859), p. 239.

—— [3] Über die Wärmeleitung gasförmiger Körper, *Pogg. Ann.* 115 (1862), p. 1.

P. u. T. Ehrenfest [1] Zur Theorie der Entropiezunahme in der statistischen Mechanik von Gibbs, *Wien Ber.* 115² (1906), p. 89.

—— [2] Über zwei bekannte Einwände gegen Boltzmanns *H*-Theorem, *Phys. Zeitschrift* 8 (1907), p. 311.

A. Einstein [1] Kinetische Theorie des Wärmegleichgewichts

und des zweiten Hauptsatzes der Thermodynamik, *Ann. d. Phys.* 9 (1902), p. 417.

—— [2] Theorie der Grundlagen der Thermodynamik, *Ann. d. Phys.* 11 (1903), p. 170.

—— [3] Über die von der molekular-kinetischen Theorie der Wärme geforderte Bewegung von in ruhenden Flüssigkeiten suspendierten Teilchen, *Ann. d. Phys.* 17 (1905), p. 549.

—— [4] Zur Theorie der Brownschen Bewegung, *Ann. d. Phys.* 19 (1906), p. 371.

J. Hadamard [1] La mécanique statistique, *Amer. Math. Soc. Bull.* (2) 12 (1906), p. 194.

G. H. Jeans [1] On the conditions necessary for equipartition of energy, *Phil. mag.* (6) 4 (1902), p. 585.

—— [2] Kinetic theory of gases, developed from a new standpoint, *Phil. mag.* (6) 5 (1903), p. 597.

A. Krönig [1] Grundzüge einer Theorie der Gase, *Pogg. Ann.* 99 (1856), p. 315.

A. Liénard [1] Sur la théorie cinétique des gaz, *J. d. phys.* (4) 2 (1903), p. 677.

G. Lippmann [1] *La théorie cinétique des gaz et le principe de Carnot, Rapports présentés au Congrès intern. de Physique réuni à Paris en 1900,* 1, Paris 1900, p. 546.

H. A. Lorentz [1] Über die Grösse von Gebieten in einer *n*-fachen Mannigfaltigkeit, *Ges. Abh.* 1, Leipzig 1906, p. 151.

—— [2] Über den zweiten Hauptsatz der Thermodynamik und dessen Beziehungen zur Molekulartheorie, *Ges. Abh.* 1, p. 202.

J. Loschmidt [1] Über das Wärmegleichgewicht eines Systems von Körpern mit Rücksicht auf die Schwere, *Wien Ber.* 73^2 (1876), p. 139; 75^2 (1877), p. 67.

J. Cl. Maxwell [1] Illustrations of the dynamical theory of gases, Part 1: On the motions and collisions of perfectly elastic spheres, *Phil. mag.* (4) 19 (1860), p. 19; Part 2: On the process of diffusions of two or more kinds of moving particles among one another, *Phil. Mag.* (4) 20 (1860), p. 1 (also *Scientific Papers* 1, Cambridge 1890, p. 377).

A CATALOG OF SELECTED

DOVER BOOKS

IN SCIENCE AND MATHEMATICS

A CATALOG OF SELECTED

DOVER BOOKS

IN SCIENCE AND MATHEMATICS

QUALITATIVE THEORY OF DIFFERENTIAL EQUATIONS, V.V. Nemytskii and V.V. Stepanov. Classic graduate-level text by two prominent Soviet mathematicians covers classical differential equations as well as topological dynamics and erqodic theory. Bibliographies. 523pp. 5⅜ × 8½. 65954-2 Pa. $10.95

MATRICES AND LINEAR ALGEBRA, Hans Schneider and George Phillip Barker. Basic textbook covers theory of matrices and its applications to systems of linear equations and related topics such as determinants, eigenvalues and differential equations. Numerous exercises. 432pp. 5⅜ × 8½. 66014-1 Pa. $8.95

QUANTUM THEORY, David Bohm. This advanced undergraduate-level text presents the quantum theory in terms of qualitative and imaginative concepts, followed by specific applications worked out in mathematical detail. Preface. Index. 655pp. 5⅜ × 8½. 65969-0 Pa. $10.95

ATOMIC PHYSICS (8th edition), Max Born. Nobel laureate's lucid treatment of kinetic theory of gases, elementary particles, nuclear atom, wave-corpuscles, atomic structure and spectral lines, much more. Over 40 appendices, bibliography. 495pp. 5⅜ × 8½. 65984-4 Pa. $11.95

ELECTRONIC STRUCTURE AND THE PROPERTIES OF SOLIDS: The Physics of the Chemical Bond, Walter A. Harrison. Innovative text offers basic understanding of the electronic structure of covalent and ionic solids, simple metals, transition metals and their compounds. Problems. 1980 edition. 582pp. 6⅛ × 9¼. 66021-4 Pa. $14.95

BOUNDARY VALUE PROBLEMS OF HEAT CONDUCTION, M. Necati Özisik. Systematic, comprehensive treatment of modern mathematical methods of solving problems in heat conduction and diffusion. Numerous examples and problems. Selected references. Appendices. 505pp. 5⅜ × 8½. 65990-9 Pa. $11.95

A SHORT HISTORY OF CHEMISTRY (3rd edition), J.R. Partington. Classic exposition explores origins of chemistry, alchemy, early medical chemistry, nature of atmosphere, theory of valency, laws and structure of atomic theory, much more. 428pp. 5⅜ × 8½. (Available in U.S. only) 65977-1 Pa. $10.95

A HISTORY OF ASTRONOMY, A. Pannekoek. Well-balanced, carefully reasoned study covers such topics as Ptolemaic theory, work of Copernicus, Kepler, Newton, Eddington's work on stars, much more. Illustrated. References. 521pp. 5⅜ × 8½. 65994-1 Pa. $11.95

PRINCIPLES OF METEOROLOGICAL ANALYSIS, Walter J. Saucier. Highly respected, abundantly illustrated classic reviews atmospheric variables, hydrostatics, static stability, various analyses (scalar, cross-section, isobaric, isentropic, more). For intermediate meteorology students. 454pp. 6⅛ × 9¼. 65979-8 Pa. $12.95

CATALOG OF DOVER BOOKS

RELATIVITY, THERMODYNAMICS AND COSMOLOGY, Richard C. Tolman. Landmark study extends thermodynamics to special, general relativity; also applications of relativistic mechanics, thermodynamics to cosmological models. 501pp. 5⅜ × 8½. 65383-8 Pa. $11.95

APPLIED ANALYSIS, Cornelius Lanczos. Classic work on analysis and design of finite processes for approximating solution of analytical problems. Algebraic equations, matrices, harmonic analysis, quadrature methods, much more. 559pp. 5⅜ × 8½. 65656-X Pa. $11.95

SPECIAL RELATIVITY FOR PHYSICISTS, G. Stephenson and C.W. Kilmister. Concise elegant account for nonspecialists. Lorentz transformation, optical and dynamical applications, more. Bibliography. 108pp. 5⅜ × 8½. 65519-9 Pa. $3.95

INTRODUCTION TO ANALYSIS, Maxwell Rosenlicht. Unusually clear, accessible coverage of set theory, real number system, metric spaces, continuous functions, Riemann integration, multiple integrals, more. Wide range of problems. Undergraduate level. Bibliography. 254pp. 5⅜ × 8½. 65038-3 Pa. $7.00

INTRODUCTION TO QUANTUM MECHANICS With Applications to Chemistry, Linus Pauling & E. Bright Wilson, Jr. Classic undergraduate text by Nobel Prize winner applies quantum mechanics to chemical and physical problems. Numerous tables and figures enhance the text. Chapter bibliographies. Appendices. Index. 468pp. 5⅜ × 8½. 64871-0 Pa. $9.95

ASYMPTOTIC EXPANSIONS OF INTEGRALS, Norman Bleistein & Richard A. Handelsman. Best introduction to important field with applications in a variety of scientific disciplines. New preface. Problems. Diagrams. Tables. Bibliography. Index. 448pp. 5⅜ × 8½. 65082-0 Pa. $10.95

MATHEMATICS APPLIED TO CONTINUUM MECHANICS, Lee A. Segel. Analyzes models of fluid flow and solid deformation. For upper-level math, science and engineering students. 608pp. 5⅜ × 8½. 65369-2 Pa. $12.95

ELEMENTS OF REAL ANALYSIS, David A. Sprecher. Classic text covers fundamental concepts, real number system, point sets, functions of a real variable, Fourier series, much more. Over 500 exercises. 352pp. 5⅜ × 8½. 65385-4 Pa. $8.95

PHYSICAL PRINCIPLES OF THE QUANTUM THEORY, Werner Heisenberg. Nobel Laureate discusses quantum theory, uncertainty, wave mechanics, work of Dirac, Schroedinger, Compton, Wilson, Einstein, etc. 184pp. 5⅜ × 8½. 60113-7 Pa. $4.95

INTRODUCTORY REAL ANALYSIS, A.N. Kolmogorov, S.V. Fomin. Translated by Richard A. Silverman. Self-contained, evenly paced introduction to real and functional analysis. Some 350 problems. 403pp. 5⅜ × 8½. 61226-0 Pa. $7.95

PROBLEMS AND SOLUTIONS IN QUANTUM CHEMISTRY AND PHYSICS, Charles S. Johnson, Jr. and Lee G. Pedersen. Unusually varied problems, detailed solutions in coverage of quantum mechanics, wave mechanics, angular momentum, molecular spectroscopy, scattering theory, more. 280 problems plus 139 supplementary exercises. 430pp. 6½ × 9¼. 65236-X Pa. $10.95

ASYMPTOTIC METHODS IN ANALYSIS, N.G. de Bruijn. An inexpensive, comprehensive guide to asymptotic methods—the pioneering work that teaches by explaining worked examples in detail. Index. 224pp. 5⅜ × 8½. 64221-6 Pa. $5.95

OPTICAL RESONANCE AND TWO-LEVEL ATOMS, L. Allen and J.H. Eberly. Clear, comprehensive introduction to basic principles behind all quantum optical resonance phenomena. 53 illustrations. Preface. Index. 256pp. 5⅜ × 8½. 65533-4 Pa. $6.95

COMPLEX VARIABLES, Francis J. Flanigan. Unusual approach, delaying complex algebra till harmonic functions have been analyzed from real variable viewpoint. Includes problems with answers. 364pp. 5⅜ × 8½. 61388-7 Pa. $7.95

ATOMIC SPECTRA AND ATOMIC STRUCTURE, Gerhard Herzberg. One of best introductions; especially for specialist in other fields. Treatment is physical rather than mathematical. 80 illustrations. 257pp. 5⅜ × 8½. 60115-3 Pa. $4.95

APPLIED COMPLEX VARIABLES, John W. Dettman. Step-by-step coverage of fundamentals of analytic function theory—plus lucid exposition of 5 important applications: Potential Theory; Ordinary Differential Equations; Fourier Transforms; Laplace Transforms; Asymptotic Expansions. 66 figures. Exercises at chapter ends. 512pp. 5⅜ × 8½. 64670-X Pa. $10.95

ULTRASONIC ABSORPTION: An Introduction to the Theory of Sound Absorption and Dispersion in Gases, Liquids and Solids, A.B. Bhatia. Standard reference in the field provides a clear, systematically organized introductory review of fundamental concepts for advanced graduate students, research workers. Numerous diagrams. Bibliography. 440pp. 5⅜ × 8½. 64917-2 Pa. $8.95

UNBOUNDED LINEAR OPERATORS: Theory and Applications, Seymour Goldberg. Classic presents systematic treatment of the theory of unbounded linear operators in normed linear spaces with applications to differential equations. Bibliography. 199pp. 5⅜ × 8½. 64830-3 Pa. $7.00

LIGHT SCATTERING BY SMALL PARTICLES, H.C. van de Hulst. Comprehensive treatment including full range of useful approximation methods for researchers in chemistry, meteorology and astronomy. 44 illustrations. 470pp. 5⅜ × 8½. 64228-3 Pa. $9.95

CONFORMAL MAPPING ON RIEMANN SURFACES, Harvey Cohn. Lucid, insightful book presents ideal coverage of subject. 334 exercises make book perfect for self-study. 55 figures. 352pp. 5⅜ × 8¼. 64025-6 Pa. $8.95

OPTICKS, Sir Isaac Newton. Newton's own experiments with spectroscopy, colors, lenses, reflection, refraction, etc., in language the layman can follow. Foreword by Albert Einstein. 532pp. 5⅜ × 8½. 60205-2 Pa. $8.95

GENERALIZED INTEGRAL TRANSFORMATIONS, A.H. Zemanian. Graduate-level study of recent generalizations of the Laplace, Mellin, Hankel, K. Weierstrass, convolution and other simple transformations. Bibliography. 320pp. 5⅜ × 8½. 65375-7 Pa. $7.95

THE ELECTROMAGNETIC FIELD, Albert Shadowitz. Comprehensive undergraduate text covers basics of electric and magnetic fields, builds up to electromagnetic theory. Also related topics, including relativity. Over 900 problems. 768pp. 5⅜ × 8¼. 65660-8 Pa. $15.95

FOURIER SERIES, Georgi P. Tolstov. Translated by Richard A. Silverman. A valuable addition to the literature on the subject, moving clearly from subject to subject and theorem to theorem. 107 problems, answers. 336pp. 5⅜ × 8½. 63317-9 Pa. $7.95

THEORY OF ELECTROMAGNETIC WAVE PROPAGATION, Charles Herach Papas. Graduate-level study discusses the Maxwell field equations, radiation from wire antennas, the Doppler effect and more. xiii + 244pp. 5⅜ × 8½. 65678-0 Pa. $6.95

DISTRIBUTION THEORY AND TRANSFORM ANALYSIS: An Introduction to Generalized Functions, with Applications, A.H. Zemanian. Provides basics of distribution theory, describes generalized Fourier and Laplace transformations. Numerous problems. 384pp. 5⅜ × 8½. 65479-6 Pa. $8.95

THE PHYSICS OF WAVES, William C. Elmore and Mark A. Heald. Unique overview of classical wave theory. Acoustics, optics, electromagnetic radiation, more. Ideal as classroom text or for self-study. Problems. 477pp. 5⅜ × 8½. 64926-1 Pa. $10.95

CALCULUS OF VARIATIONS WITH APPLICATIONS, George M. Ewing. Applications-oriented introduction to variational theory develops insight and promotes understanding of specialized books, research papers. Suitable for advanced undergraduate/graduate students as primary, supplementary text. 352pp. 5⅜ × 8½. 64856-7 Pa. $8.50

A TREATISE ON ELECTRICITY AND MAGNETISM, James Clerk Maxwell. Important foundation work of modern physics. Brings to final form Maxwell's theory of electromagnetism and rigorously derives his general equations of field theory. 1,084pp. 5⅜ × 8½. 60636-8, 60637-6 Pa., Two-vol. set $19.00

AN INTRODUCTION TO THE CALCULUS OF VARIATIONS, Charles Fox. Graduate-level text covers variations of an integral, isoperimetrical problems, least action, special relativity, approximations, more. References. 279pp. 5⅜ × 8½. 65499-0 Pa. $6.95

HYDRODYNAMIC AND HYDROMAGNETIC STABILITY, S. Chandrasekhar. Lucid examination of the Rayleigh-Benard problem; clear coverage of the theory of instabilities causing convection. 704pp. 5⅜ × 8¼. 64071-X Pa. $12.95

CALCULUS OF VARIATIONS, Robert Weinstock. Basic introduction covering isoperimetric problems, theory of elasticity, quantum mechanics, electrostatics, etc. Exercises throughout. 326pp. 5⅜ × 8½. 63069-2 Pa. $7.95

DYNAMICS OF FLUIDS IN POROUS MEDIA, Jacob Bear. For advanced students of ground water hydrology, soil mechanics and physics, drainage and irrigation engineering and more. 335 illustrations. Exercises, with answers. 784pp. 6⅛ × 9¼. 65675-6 Pa. $19.95

ORDINARY DIFFERENTIAL EQUATIONS, Morris Tenenbaum and Harry Pollard. Exhaustive survey of ordinary differential equations for undergraduates in mathematics, engineering, science. Thorough analysis of theorems. Diagrams. Bibliography. Index. 818pp. 5⅜ × 8½. 64940-7 Pa. $15.95

STATISTICAL MECHANICS: Principles and Applications, Terrell L. Hill. Standard text covers fundamentals of statistical mechanics, applications to fluctuation theory, imperfect gases, distribution functions, more. 448pp. 5⅜ × 8½. 65390-0 Pa. $9.95

ORDINARY DIFFERENTIAL EQUATIONS AND STABILITY THEORY: An Introduction, David A. Sánchez. Brief, modern treatment. Linear equation, stability theory for autonomous and nonautonomous systems, etc. 164pp. 5⅜ × 8¼. 63828-6 Pa. $4.95

THIRTY YEARS THAT SHOOK PHYSICS: The Story of Quantum Theory, George Gamow. Lucid, accessible introduction to influential theory of energy and matter. Careful explanations of Dirac's anti-particles, Bohr's model of the atom, much more. 12 plates. Numerous drawings. 240pp. 5⅜ × 8½. 24895-X Pa. $5.95

ORDINARY DIFFERENTIAL EQUATIONS, I.G. Petrovski. Covers basic concepts, some differential equations and such aspects of the general theory as Euler lines, Arzel's theorem, Peano's existence theorem, Osgood's uniqueness theorem, more. 45 figures. Problems. Bibliography. Index. xi + 232pp. 5⅜ × 8½. 64683-1 Pa. $6.00

GREAT EXPERIMENTS IN PHYSICS: Firsthand Accounts from Galileo to Einstein, edited by Morris H. Shamos. 25 crucial discoveries: Newton's laws of motion, Chadwick's study of the neutron, Hertz on electromagnetic waves, more. Original accounts clearly annotated. 370pp. 5⅜ × 8½. 25346-5 Pa. $8.95

INTRODUCTION TO PARTIAL DIFFERENTIAL EQUATIONS WITH APPLICATIONS, E.C. Zachmanoglou and Dale W. Thoe. Essentials of partial differential equations applied to common problems in engineering and the physical sciences. Problems and answers. 416pp. 5⅜ × 8½. 65251-3 Pa. $9.95

BURNHAM'S CELESTIAL HANDBOOK, Robert Burnham, Jr. Thorough guide to the stars beyond our solar system. Exhaustive treatment. Alphabetical by constellation: Andromeda to Cetus in Vol. 1; Chamaeleon to Orion in Vol. 2; and Pavo to Vulpecula in Vol. 3. Hundreds of illustrations. Index in Vol. 3. 2,000pp. 6⅛ × 9¼. 23567-X, 23568-8, 23673-0 Pa., Three-vol. set $38.85

ASYMPTOTIC EXPANSIONS FOR ORDINARY DIFFERENTIAL EQUATIONS, Wolfgang Wasow. Outstanding text covers asymptotic power series, Jordan's canonical form, turning point problems, singular perturbations, much more. Problems. 384pp. 5⅜ × 8½. 65456-7 Pa. $8.95

AMATEUR ASTRONOMER'S HANDBOOK, J.B. Sidgwick. Timeless, comprehensive coverage of telescopes, mirrors, lenses, mountings, telescope drives, micrometers, spectroscopes, more. 189 illustrations. 576pp. 5⅜ × 8¼. 24034-7 Pa. $8.95

SPECIAL FUNCTIONS, N.N. Lebedev. Translated by Richard Silverman. Famous Russian work treating more important special functions, with applications to specific problems of physics and engineering. 38 figures. 308pp. 5⅜ × 8½.
60624-4 Pa. $6.95

OBSERVATIONAL ASTRONOMY FOR AMATEURS, J.B. Sidgwick. Mine of useful data for observation of sun, moon, planets, asteroids, aurorae, meteors, comets, variables, binaries, etc. 39 illustrations 384pp. 5⅜ × 8¼. (Available in U.S. only)
24033-9 Pa. $5.95

INTEGRAL EQUATIONS, F.G. Tricomi. Authoritative, well-written treatment of extremely useful mathematical tool with wide applications. Volterra Equations, Fredholm Equations, much more. Advanced undergraduate to graduate level. Exercises. Bibliography. 238pp. 5⅜ × 8½.
64828-1 Pa. $6.95

CELESTIAL OBJECTS FOR COMMON TELESCOPES, T.W. Webb. Inestimable aid for locating and identifying nearly 4,000 celestial objects. 77 illustrations. 645pp. 5⅜ × 8½.
20917-2, 20918-0 Pa., Two-vol. set $12.00

MODERN NONLINEAR EQUATIONS, Thomas L. Saaty. Emphasizes practical solution of problems; covers seven types of equations. ". . . a welcome contribution to the existing literature. . . ."—*Math Reviews*. 490pp. 5⅜ × 8½.
64232-1 Pa. $9.95

FUNDAMENTALS OF ASTRODYNAMICS, Roger Bate et al. Modern approach developed by U.S. Air Force Academy. Designed as a first course. Problems, exercises. Numerous illustrations. 455pp. 5⅜ × 8½.
60061-0 Pa. $8.95

INTRODUCTION TO LINEAR ALGEBRA AND DIFFERENTIAL EQUATIONS, John W. Dettman. Excellent text covers complex numbers, determinants, orthonormal bases, Laplace transforms, much more. Exercises with solutions. Undergraduate level. 416pp. 5⅜ × 8½.
65191-6 Pa. $8.95

INCOMPRESSIBLE AERODYNAMICS, edited by Bryan Thwaites. Covers theoretical and experimental treatment of the uniform flow of air and viscous fluids past two-dimensional aerofoils and three-dimensional wings; many other topics. 654pp. 5⅜ × 8½.
65465-6 Pa. $14.95

INTRODUCTION TO DIFFERENCE EQUATIONS, Samuel Goldberg. Exceptionally clear exposition of important discipline with applications to sociology, psychology, economics. Many illustrative examples; over 250 problems. 260pp. 5⅜ × 8½.
65084-7 Pa. $6.95

LAMINAR BOUNDARY LAYERS, edited by L. Rosenhead. Engineering classic covers steady boundary layers in two- and three-dimensional flow, unsteady boundary layers, stability, observational techniques, much more. 708pp. 5⅜ × 8½.
65646-2 Pa. $15.95

LECTURES ON CLASSICAL DIFFERENTIAL GEOMETRY, Second Edition, Dirk J. Struik. Excellent brief introduction covers curves, theory of surfaces, fundamental equations, geometry on a surface, conformal mapping, other topics. Problems. 240pp. 5⅜ × 8½.
65609-8 Pa. $6.95

GEOMETRY OF COMPLEX NUMBERS, Hans Schwerdtfeger. Illuminating, widely praised book on analytic geometry of circles, the Moebius transformation, and two-dimensional non-Euclidean geometries. 200pp. 5⅝ × 8¼.
63830-8 Pa. $6.95

MECHANICS, J.P. Den Hartog. A classic introductory text or refresher. Hundreds of applications and design problems illuminate fundamentals of trusses, loaded beams and cables, etc. 334 answered problems. 462pp. 5⅜ × 8½. 60754-2 Pa. $8.95

TOPOLOGY, John G. Hocking and Gail S. Young. Superb one-year course in classical topology. Topological spaces and functions, point-set topology, much more. Examples and problems. Bibliography. Index. 384pp. 5⅝ × 8¼.
65676-4 Pa. $7.95

STRENGTH OF MATERIALS, J.P. Den Hartog. Full, clear treatment of basic material (tension, torsion, bending, etc.) plus advanced material on engineering methods, applications. 350 answered problems. 323pp. 5⅜ × 8½. 60755-0 Pa. $7.50

ELEMENTARY CONCEPTS OF TOPOLOGY, Paul Alexandroff. Elegant, intuitive approach to topology from set-theoretic topology to Betti groups; how concepts of topology are useful in math and physics. 25 figures. 57pp. 5⅜ × 8½.
60747-X Pa. $2.95

ADVANCED STRENGTH OF MATERIALS, J.P. Den Hartog. Superbly written advanced text covers torsion, rotating disks, membrane stresses in shells, much more. Many problems and answers. 388pp. 5⅜ × 8½. 65407-9 Pa. $8.95

COMPUTABILITY AND UNSOLVABILITY, Martin Davis. Classic graduate-level introduction to theory of computability, usually referred to as theory of recurrent functions. New preface and appendix. 288pp. 5⅜ × 8½. 61471-9 Pa. $6.95

GENERAL CHEMISTRY, Linus Pauling. Revised 3rd edition of classic first-year text by Nobel laureate. Atomic and molecular structure, quantum mechanics, statistical mechanics, thermodynamics correlated with descriptive chemistry. Problems. 992pp. 5⅜ × 8½. 65622-5 Pa. $18.95

AN INTRODUCTION TO MATRICES, SETS AND GROUPS FOR SCIENCE STUDENTS, G. Stephenson. Concise, readable text introduces sets, groups, and most importantly, matrices to undergraduate students of physics, chemistry, and engineering. Problems. 164pp. 5⅜ × 8½. 65077-4 Pa. $5.95

THE HISTORICAL BACKGROUND OF CHEMISTRY, Henry M. Leicester. Evolution of ideas, not individual biography. Concentrates on formulation of a coherent set of chemical laws. 260pp. 5⅜ × 8½. 61053-5 Pa. $6.00

THE PHILOSOPHY OF MATHEMATICS: An Introductory Essay, Stephan Körner. Surveys the views of Plato, Aristotle, Leibniz & Kant concerning propositions and theories of applied and pure mathematics. Introduction. Two appendices. Index. 198pp. 5⅜ × 8½. 25048-2 Pa. $5.95

THE DEVELOPMENT OF MODERN CHEMISTRY, Aaron J. Ihde. Authoritative history of chemistry from ancient Greek theory to 20th-century innovation. Covers major chemists and their discoveries. 209 illustrations. 14 tables. Bibliographies. Indices. Appendices. 851pp. 5⅜ × 8½. 64235-6 Pa. $15.95

THE FOUR-COLOR PROBLEM: Assaults and Conquest, Thomas L. Saaty and Paul G. Kainen. Engrossing, comprehensive account of the century-old combinatorial topological problem, its history and solution. Bibliographies. Index. 110 figures. 228pp. 5⅜ × 8½. 65092-8 Pa. $6.00

CATALYSIS IN CHEMISTRY AND ENZYMOLOGY, William P. Jencks. Exceptionally clear coverage of mechanisms for catalysis, forces in aqueous solution, carbonyl- and acyl-group reactions, practical kinetics, more. 864pp. 5⅜ × 8½. 65460-5 Pa. $18.95

PROBABILITY: An Introduction, Samuel Goldberg. Excellent basic text covers set theory, probability theory for finite sample spaces, binomial theorem, much more. 360 problems. Bibliographies. 322pp. 5⅜ × 8½. 65252-1 Pa. $7.95

LIGHTNING, Martin A. Uman. Revised, updated edition of classic work on the physics of lightning. Phenomena, terminology, measurement, photography, spectroscopy, thunder, more. Reviews recent research. Bibliography. Indices. 320pp. 5⅜ × 8¼. 64575-4 Pa. $7.95

PROBABILITY THEORY: A Concise Course, Y.A. Rozanov. Highly readable, self-contained introduction covers combination of events, dependent events, Bernoulli trials, etc. Translation by Richard Silverman. 148pp. 5⅜ × 8¼. 63544-9 Pa. $4.50

THE CEASELESS WIND: An Introduction to the Theory of Atmospheric Motion, John A. Dutton. Acclaimed text integrates disciplines of mathematics and physics for full understanding of dynamics of atmospheric motion. Over 400 problems. Index. 97 illustrations. 640pp. 6 × 9. 65096-0 Pa. $16.95

STATISTICS MANUAL, Edwin L. Crow, et al. Comprehensive, practical collection of classical and modern methods prepared by U.S. Naval Ordnance Test Station. Stress on use. Basics of statistics assumed. 288pp. 5⅜ × 8½. 60599-X Pa. $6.00

WIND WAVES: Their Generation and Propagation on the Ocean Surface, Blair Kinsman. Classic of oceanography offers detailed discussion of stochastic processes and power spectral analysis that revolutionized ocean wave theory. Rigorous, lucid. 676pp. 5⅜ × 8½. 64652-1 Pa. $14.95

STATISTICAL METHOD FROM THE VIEWPOINT OF QUALITY CONTROL, Walter A. Shewhart. Important text explains regulation of variables, uses of statistical control to achieve quality control in industry, agriculture, other areas. 192pp. 5⅜ × 8½. 65232-7 Pa. $6.00

THE INTERPRETATION OF GEOLOGICAL PHASE DIAGRAMS, Ernest G. Ehlers. Clear, concise text emphasizes diagrams of systems under fluid or containing pressure; also coverage of complex binary systems, hydrothermal melting, more. 288pp. 6½ × 9¼. 65389-7 Pa. $8.95

STATISTICAL ADJUSTMENT OF DATA, W. Edwards Deming. Introduction to basic concepts of statistics, curve fitting, least squares solution, conditions without parameter, conditions containing parameters. 26 exercises worked out. 271pp. 5⅜ × 8½. 64685-8 Pa. $7.95

NOTES FROM INDOCHINA

on ethnic minority cultures

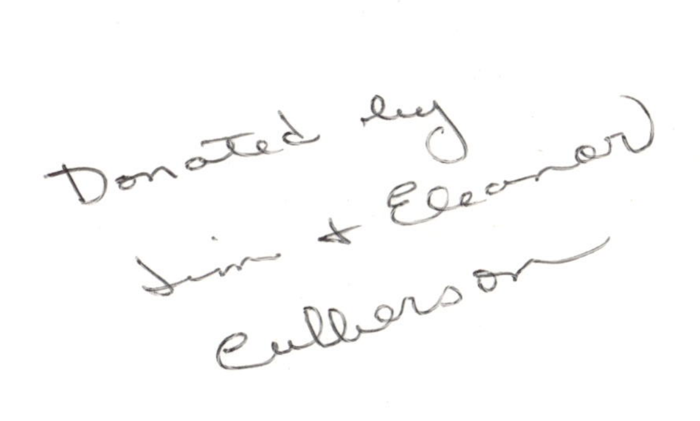

SIL MUSEUM OF ANTHROPOLOGY

PUBLICATION 6